HILL FARMING

Edited by
Hazelle S. Cabugao

ArclerPress

Hill Farming

Edited by Hazelle S. Cabugao

Edition 2018

ArclerPress

Published by

Arcler Press LLC

708 3rd Avenue, 6th Floor

New York

NY 10017

United States of America

www.arclerpress.com

ISBN: 978-1-68094-565-2 (Hardback)

Printed in United States of America

Library of Congress Control Number: 2017933135

About the Editor

Hazelle S. Cabugao

Hazelle earned her degree in Agriculture from the University of the Philippines Los Baños in 2014. Her studies focus on ornamental horticulture particularly the use of plants and trees in designing landscapes. A horticulturist by profession, she supervises landscape implementation and maintenance and is also passionate about conserving the indigenous flora

Table of Contents

3 MOORLAND

7 RICE FARMING IN UPLAND AREAS

8 SOIL AND WATER CONSERVATION THROUGH HORTICULTURAL INTERVENSION IN HILLY AREAS

List of Abbreviations

AONB	Area of Outstanding Natural Beauty
CCC	Civilian Conservation Corps
CAP	Common Agricultural Policy
CSS	Context Sensitive Solutions
DAs	Disadvantaged Areas
FTA	Federal Transit Administration
HFA	Hill Farming Allowance
IBA	Important Bird Area
KAIST	Korea Advanced Institute of Science and Technology
LFA	Less Favoured Area
Mg	Magnesium
Mn	Manganese
MPO	Metropolitan Planning Organization
MoD	Ministry of Defence
Mb	Molybdenum
SPS	Single Payment Scheme
SAC	Special Areas of Conservation
PDNP	Peak District National Park
RDPE	Rural Development Programme for England
UNEP	United Nations Environment Programme
UTMS	Urban Transportation Modeling System
SDAs	Severely Disadvantaged Areas

Preface

Hill farming is extensive farming in upland areas, primarily rearing sheep, although historically cattle were often reared extensively in upland areas. Hill farming is a type of agricultural practice in the UK in upland regions. In England, hill farms are located mainly in the North and South-Western regions, as well as a few areas bordering Wales. The Scottish highlands are another home for many hill farms. Sheep farms and mixed sheep and cattle farms constitute approximately 55% of the agricultural land in Scotland. These areas have a harsh climate, short growing seasons, relatively poor quality of soil and long winters. Therefore, these areas are considered to be disadvantaged and the animals raised there are generally less productive and farmers will often send them down to the lowlands to be fattened up. Upland areas are not traditionally favorable for agricultural practices. Sustainable farming systems in upland areas are one of the greatest challenges facing agriculture, since a balance is sought between economic development and environmental protection in those areas. Uplands are particularly sensitive to agricultural encroachments. Driven by growing food demand to feed increasing populations and low farm income in many uplands, however, there is a tendency to use more productive, intensive farming methods in place of traditional subsistence farming characterized by poor crop yields and low farm productivity. Intensive farming methods suitable for lowlands can be disastrous when used on uplands without proven technologies and experience, promoting deforestation and soil erosion and reducing land productivity. The problem of sustainable upland agriculture is not a technical one as such but it is more institutional, involving limited R&D investment in upland farming research, sociopolitical neglect of marginalized upland societies, low capacity of communities, and inappropriate development planning. In recent years, there have been some successful examples of sustainable upland farming. Farmers need cash crops as well as food crops, in systems which maintain soil productivity. Where transport networks are poor, low-volume high-value cash crops with a long storage life are important. Many farming households now earn a living from direct marketing of specialty crops and animals, accommodation, or recreation and leisure, as well as farming.

Content Coverage

Chapter One presents an introduction of hill farming principles and practices.

Chapter Two presents an overview of fell, such as a mountain range or moor-covered hills.

Chapter Three focuses on moorland, a type of habitat found in upland areas in temperate grasslands, savannas, and shrublands and montane grasslands and shrublands biomes, characterised by low-growing vegetation on acidic soils. Moorland nowadays generally means uncultivated hill land, but includes low-lying wetlands.

Chapter Four focuses on pastoral farming, in which the livestock farmer has some form of ownership of the land used, giving the farmer more economic incentive to improve the land.

Chapter Five is concerned with contour plowing that follows the curves of the landscape instead of plowing straight rows up and down hills and valleys.

Chapter Six aims to provide agricultural extensionists with basic information which will help them design plans for the conservation of soils and the management of water runoff in specific agricultural plots.

Chapter Seven focuses on the practices related to rice farming in upland areas. New challenges are emerging in the world's upland rice farming areas, where already some of the world's poorest farmers try to wrest a living from fragile soils that are fast being degraded.

Chapter Eight identifies the conservation of land resources that can promote sound land use to match with the land capabilities or suitability and to initiate correct land resources.

CHAPTER 1

Hill Farming

Chapter Contents

Hill farming uses specific hardy breeds of sheep and cattle that are adapted to a harsh climate, long winters and poor pastures. These animals are typically less productive and will be sold as store stock to be fattened and finished in lowland farms.

INTRODUCTION

Hill farming is extensive farming in upland areas, primarily rearing sheep, although historically cattle were often reared extensively in upland areas. Fell farming is the farming of fells, a fell being an area of uncultivated high ground used as common grazing. It is a term commonly used in Northern England, especially in the Lake District and the Pennine Dales. Elsewhere, the terms *hill farming* or pastoral farming are more commonly used.

Cattle farming in the hills is usually restricted by a scarcity of winter fodder, and hill sheep, grazing at about two hectares per head, are often taken to lowland areas for fattening.

Modern hill farming is often heavily dependent on state subsidy, for example in the United Kingdom it receives support from the European Union›s Common Agricultural

Policy. Improved, sown pasture and drained moorland can be stocked more heavily, at approximately one sheep per 0.25 hectares.

An example of hill farming countryside in the UK

HILL FARMING LOCATION AND ORGANIZATION

Relief map of the United Kingdom

Hill farming is a type of agricultural practice in the UK in upland regions. In England, hill farms are located mainly in the North and South-Western regions, as well as a few areas bordering Wales. The Scottish highlands are another home for many hill farms. Sheep farms and mixed sheep and cattle farms constitute approximately 55% of the agricultural land in Scotland. These areas have a harsh climate, short growing seasons, relatively poor quality of soil and long winters. Therefore, these areas are considered to be disadvantaged and the animals raised there are generally less productive and farmers will often send them down to the lowlands to be fattened up.

Upland areas are not traditionally favourable for agricultural practices. The majority of Hill farming land in England is classified as Less Favoured Area (LFA), and the LFA constitutes 17% of land farmed in England. The LFA is further divided into Severely Disadvantaged Areas (SDAs) and Disadvantaged Areas (DAs), which make up 67% and 33% of the LFA respectively. These areas are classified as such on account of poor climate, soils, and terrain which cause higher costs in production and transportation as well as lower yields and less productivity. The LFA is significant in England's farming on a whole despite these disadvantages: 30% of beef cows and 44% of breeding sheep come from LFAs. Farming distinctively shapes the ecosystems of these zones, and the agriculture practices in the uplands define and shape the environment and landscape.

Upland areas are usually covered with both dry and wet dwarf shrub heath and, rough and either managed or unmanaged improved grasslands. The typical hill farm is made up of three distinct zones: the High fell, the Alotment, and the Inbye. The High fell includes peat moors and rocky areas which provide poor grazing at the top. The Alotment follows below, an enclosed area with rough grazing. The Inbye is the lowest area at the bottom, which is used as the regular grazing area as well as for growing hay.

HILL FARMING THROUGHOUT HISTORY

Dartmoor National Park has over 10,000ha of prehistoric field systems, dating back to 1500BC. Archaeological evidence shows that these moors have been grazed for 3500 years. Because of the extent of historical farming in the UK uplands, hill farming has shaped the English uplands both environmentally and culturally.

The UK government has designated different areas in the upland as specifically valuable by certain terms of the environment, biodiversity, archaeology, cultural heritage and landscape, and seeks to protect these regions for such reasons. According to the UK government, these designated areas and their qualifications as such are:

- National parks
    - "These protect and conserve the character of landscapes, facilities for access, wildlife habitats and historic features"
- Areas of outstanding natural beauty
    - "These conserve and enhance naturally beautiful landscapes"
- Countryside character areas
    - "Areas of cultural heritage which should be preserved"
- Natural England natural areas

  - "Each area has a unique identity created by its mix of natural features and human activities and provides a broad context for local nature conservation work"
- Common land
  - "Areas where people who do not own the land have rights to use it for farming or other purposes"
- National nature reserves
  - "These protect and provide public access to important wildlife and geological sites"
- Special areas of conservation
  - "These protect various wild animals, plants and habitats under the European Union's Habitats Directive"
- Special protection areas
  - "These protect rare and vulnerable birds and migratory species as well as geological and physiographical heritage"
- Upland experiment areas
  - "Two upland areas where Natural England/Defra predecessor bodies piloted an integrated approach to rural development and nature conservation between 1999 and 2001"
- Ancient woodland
  - "Land that has had continuous tree cover since at least 1600"

Over the past century, Hill farming and the upland environment have undergone a number of changes. Since 1900 there has been:

- An approximately 500% increase in the number of sheep livestock
- A decrease in medium-sized farms, due to increase in large farms businesses and the emergence of small-scale hobby farmers.
- Increased specialization in livestock and a movement away from tradition mixed farming methods.
- Fairly consistent labour employment on account of constant agricultural intensification offsetting reductions in labour output made possible by technological advancements.
- A high turnover rates in upland ecosystem habitat types. For example, although the percentage of land classified as dwarf shrub moor remained relatively stable between 1913 and 2000, only 55% of the dwarf moor shrub land in 1913 occupied the same area as it does in 2000.

HILL FARMING AND UPLANDS ECOSYSTEMS

A large number of upland ecosystems have been shaped by humans for centuries, particularly by farming and agriculture. Because of this, many upland ecosystems have become dependent on hill farm land management. Hill farming practices play a significant role in supporting surrounding flora and fauna in the uplands. Through grazing, sheep and cattle maintain a variety of tall grasses and short vegetation. This in turn supports local wildlife, as the short vegetation provides breeding and nesting grounds for many species of waders, including the lapwing, redshank, and golden plover. The taller grasses are an important part of the Curlew habitat, which is another species of wader. Cattle dung provides nutrition for many species of insects and carrion provides food for various species of scavenging birds.

The Dartmoor sheep is a type of livestock found on hill farms

During winter farmers will usually keep the animals indoors, supplementing the livestock›s diet with hay or silage. The land used to grow winter feed that are not mowed are able to provide protection for a variety of birds including skylarks, partridge, and corncrakes who build on their nests on the ground. Agricultural use, burning, and grazing by both livestock and wild life such as deer, helps to sustain the upland grasslands, moorland and bogs. If these ecosystems were not maintained they would be colonized by trees and scrub.

Sustainable careful maintenance is highly important in hill farming in order to protect the delicate relationship that farm manage has on the biodiversity of native plant and animal species.

A Grey Faced Dartmoor sheep

Upland ecosystems have seen a shift in the last century, associated with widespread habitat deterioration caused by human actions and exploitation. The decline in grazing animals accompanied with the milder winters experienced in recent years has caused an overgrowth in vegetation, putting the ecosystem, as well as various archaeological sites at risk. The Dartmoor Vision initiative is trying to return Dartmoor to its former predominantly cattle, sheep, and pony grazed landscape.

GOVERNMENT SUPPORT AND SUBSIDIES

Hill farm incomes in the UK have recently seen great decrease following drops in lamb and beef prices.Therefore, subsidy support has become vital for Hill farm survival, and the policies have been changing in response to continuous uncertainty in the sector.

Common Agricultural Policy (CAP)

Hill farming has been supported by both the British government and EU policies, one of the most influential EU scheme being The Common Agricultural Policy (CAP). The CAP provided production-based direct (headage) which gave incentive to stock beef cattle and sheep at high densities. This led to, in some circumstances, overgrazing which damages natural and semi-natural vegetation. Because of overgrazing and issues with the accumulation of surpluses, the CAP was reformed. The two most recent reforms to the CAP were Agenda 2000 in 1999 and the Mid Term Review of June 2003 and April

2004. These changes are phasing out support and protections linked to production, and are providing more support on environmental and rural developments.

Single Farm Payment

The Single Farm Payment replaced the older headage payments (CAP) in 2005. Analyses of the effects of economic incentives provided to hill farmers by decoupling and the introduction of the Single Farm Payment show that although these policies cause little change in average farm incomes they do encourage change in the way hill farms run. Specifically the policies promote the reduction of stocking densities, reduction of employment of additional farm labour, movement away from reliance on beef cattle, increased specialization, and to keep farming land in "good agricultural condition" rather than farm abandonment. The EU plans to phase out and progressively reduce the SFP, and the SFP is guaranteed until 2013.

Other Policies

Other subsidy schemes from the British government are available to hill farmers, particularly the Uplands Entry Level Stewardship (Uplands ELS)and agri-environment schemes. The Uplands ELS replaced the prior Hill Farm Allowance in 2010. Before the HFA, hill farms we subsidized by the Hill Livestock Compensatory Allowances which were active as headage support to eligible beef cows and ewes. Because the DAs are more profitable than the SDAs, active since 2008 DAs became ineligible for funding from the Hill Farming Allowance (HFA).

In addition to the Upland ELS, hill farmers in England's SDAs are supported by the Single Payment Scheme (SPS), which is the primary agricultural subsidy scheme under the EU. Subsidies from the SPS are not dependent on production, granting greater freedom to farmers to meet market demands. The SPS also claims to specifically support hill farmers who follow environmentally friendly farming practices.

In order to receive these subsidies, hill farmers must meet cross compliance rules and regulations, which mainly involves avoidance of overgrazing and unsuitable supplementary feeding on natural and semi-natural vegetation under GAEC (standard of good agricultural and environmental condition). These standards were implemented to protect significant habitats and to limit soil erosion and other negative effects of soil structure in the uplands.

Certain upland farmers and communities also have access to funding from the Rural Development Programme for England (RDPE) team at Defra. The reduction of farming subsidies that have taken place over the past few decades has created an uncertain future for farming in many parts of Europe.

Recent Strain on Hill Farming

Hill farming countryside by the Arkland Burn

Without government subsidies, many hill farms would have a negative income. The high cost of land and machinery keeps many hill farmers from expanding. Hill farmers in some parts of England have reported a 75% decrease in income over the past decade.

Hill farmer income is subject to large fluctuation above the influence of the farmers. The harsh terrain and climate of hill farms are hard on the animals, causing them to be relatively very unproductive. Because of this, hill farming can have economic strains on the farmers who generally have low income. Wet weather, as often experienced in the uplands, create additional animal feed costs for farmers.

Many hill farmers earn around £12,600, with some earning as little as £8,000. This is much below the annual £19,820 a single working adult requires to live in a village in England. In 2008, a farmer would receive a profit of £1 for a single moorland lamb. The average LFA farm in England only earns about 66% of their total revenue from farming. 22% of this revenue comes from the Single Farm Payment, and 10% from specific agri-environment payments. The 2% balance originated from non-farm activities, which are usually associated with contracting or tourism and recreation.

Hill farmers in Peak District National Park (PDNP) constitute one of UK's most deprived farming communities, with farms in the LFA making an average loss of £16,000 per farm, generating an average headline Farm Business Income of £10,800 (supplemented by various government subsidies), creating a net income average per farm of about £6000.

The hill farming sector in southwest England, like farming in the rest of the country, has experienced a decade of much change associated with economic pressures and uncertainties. On average, the financial position of hill farms in South West England, like the rest of the country, is precarious. The average southwest English hill farm system in unable to match labour and capital invest in the business.

Many farmers rely on a Single Farm Payment as a source of income. These payments are expected to arrive in November or December, but sometimes farmers do not receive the money until June. Due to this farmers are often unable to pay their bills or fix their machinery. Some farmers have to cut back on the feed given to their animals, leading to a decrease in meat production and therefore lower profit. By 2012 the Single Payment Scheme (or SPS), will only take into consideration the area of the farm. This will decrease the income in moorland farmers to only 70% of what it was 20 years ago.

The income from calves and lambs has remained constant, while the costs of farm upkeep have risen sharply (including items such as feed, straw, fuel, or fertilizer). Because hill farming is becoming increasingly less profitable an increasing number of farmers have switched from the traditional hearty but less profitable animals which graze the moors to mainstream more profitable animals.

Opportunities for farmers to supplement their farm income by working in industries such as quarrying or mining are largely no longer available.The financial burden has taken a toll on many hill farmers, causing them to exhibit signs of mental health issues. Many hill farmers are forced to generate supplemental income outside their farms or to take out loans. Because of these economic factors, there is little incentive for younger generations to continue on with the hill farming.

Problems

As discussed in an article on the Carnegie UK Trust Rural Community Development Programme site:

"The Foot and Mouth outbreak in Cumbria in 2001 led to the culling of over a million sheep. It also showed that **the hill (fell) farming communities** were as vulnerable as the pastoral system they have created over many generations."

REFERENCES

1. Hill farming - Detailed guidance - GOV.UK
2. Hill farming - Scottish Natural Heritage
3. khttp://www.fbspartnership.co.uk/documents/2010_11/Hill_Farming_in_England_10-11.pdf
4. BBC - Standard Grade Bitesize Geography - Hill farming : Revision
5. http://www.dartmoor-npa.gov.uk/__data/assets/pdf_file/0007/42010/dartmoor_vision_leaflet_april_2008.pdf
6. Armsworth 2009
7. ihttp://www.esds.ac.uk/doc/6363/mrdoc/pdf/6363uguide.pdf
8. Hard Times: A research report into hill farming and farming families in the Peak District
9. ihttp://oxfamilibrary.openrepository.com/oxfam/bitstream/10546/249113/1/rr-challenges-facing-farmers-in-teesdale-121012-en.pdf
10. http://swuf.org.uk/wp-content/uploads/2012/07/Hill-farming-systems-in-SW-report-2008.pdf
11. Rural Community Development Programme - Media Clips

CHAPTER 2

The Concept of Fell

Chapter Contents

- **Introduction**
- **Etymology**
- **Fennoscandia**

INTRODUCTION

A fell (from Old Norse fell, fjall, "mountain") is a high and barren landscape feature, such as a mountain range or moor-covered hills. The term is most often employed in Fennoscandia, the Isle of Man, parts of Northern England, and Scotland.

Fjäll landscape in Padjelanta, Swedish Lapland

The Ylläs fell in Kolari, Finnish Lapland

Børvasstindene in Norway, near Bodø

ETYMOLOGY

The English word *fell* comes from Old Norse *fell, fjall* (both forms existed). It is cognate with Icelandic *fjall/fell*, Faroese *fjall*, Danish *fjeld*, Swedish *fjäll*, and Norwegian*fjell*, all referring to mountains rising above the alpine tree line.

British Isles

View of the Scafell massif from Yewbarrow, Wasdale, Cumbria. In the valley are older enclosures and higher up on the fell-side are the parliamentary enclosures following straight lines regardless of terrain.

In Northern England, especially in the Lake District and in the Pennine Dales, the word "fell" originally referred to an area of uncultivated high ground used as common grazing usually on common land and above the timberline. Today, generally, "fell" refers to the mountains and hills of the Lake District and the Pennine Dales.

Names that originally referred to grazing areas have been applied to these hilltops. This is the case with Seathwaite Fell, for example, which would be the common grazing land used by the farmers of Seathwaite. The fellgate marks the road from a settlement onto the fell (see photograph for example), as is the case with the Seathwaite Fell. In other cases the reverse is true; for instance, the name of Wetherlam, in the Coniston Fells, though understood to refer to the mountain as a whole, strictly speaking refers to the summit; the slopes have names such as Tilberthwaite High Fell, Low Fell and Above Beck Fells.

Fell Lane, near Ingleton towards the fellgate and Ingleborough, North Yorkshire, England

The word fell is also used in the names of various breeds of livestock, bred for life on the uplands, such as Rough Fell sheep and Fell ponies.

It is also found in many place names across the North of England, often attached to the name of a community; thus the township of Cartmel Fell.

In northern England, there is a Lord of the Fells – this ancient aristocratic title being associated with the Lords of Bowland.

Groups of cairns are a common feature on many fells, often marking the summit – there are fine examples on Wild Boar Fell in Mallerstang Dale, Cumbria, and on Nine Standards Rigg just outside Kirkby Stephen, Cumbria.

As the most mountainous region of England, the Lake District is the area most closely associated with the sport of fell running, which takes its name from the fells of the district. Fellwalking is also the term used locally for the activity known in the rest of Great Britain as hillwalking.

The word fell also enjoys limited use in Scotland, with for example the Campsie Fells in Central Scotland, to the North East of Glasgow. One of the most famous examples of the use of the word fell in Scotland is Goat Fell, the highest point on the Isle of Arran. Criffel and the nearby Long Fell in Galloway may be seen from the northern Lake District of England. Peel Fell in the Kielder Forest is situated on the border between the Scottish Borders to the North and the English county of Northumberland to the South.

FENNOSCANDIA

In Norway, *fjell*, in common usage, is generally interpreted as simply a summit of greater altitude than a hill, which leads to a great deal of local variation in what is defined as a 'fjell'. Professor of geography at the University of Bergen, Anders Lundeberg, has summed up the problem by stating that "There simply is no fixed and unambiguous definition of 'fjell'." Ivar Aasen defined *fjell* as a "tall berg", primarily referring to a *berg* that reaches an altitude where trees don't grow, lower *berg* are referred to as "berg", "ås (hill, ridge) or hei (moor, heathland). The fixed expression *til fjells* refers to mountains (or uplands) as a collective rather than a specific location or specific summit. According to Ivar Aasen *berg* refers to cliffs, bedrock and notable elevations of the surface underpinned by bedrock; *berg* also refers to the substance of bedrock. For all practical purposes, 'fjell› can be translated as ‹mountain› and the Norwegian language has no other commonly used word for mountain.

In Sweden, fjäll refers to any mountain or upland high enough that forest will not naturally survive at the top, in effect a mountain tundra. ‹Fjäll› is primarily used to describe mountains in the Nordic countries, but also more generally to describe mountains shaped by massive ice sheets, primarily in Arctic and subarctic regions.

In Finnish, the mountains characteristic of the region of Lapland are called *tunturi* (plural: *tunturit*). In Finnish, the geographical term *vuori* is used for mountains recently uplifted and with jagged terrain featuring permanent glaciers, while *tunturi* refers to the old, highly eroded, gently shaped terrain without glaciers, as found in Finland. (Also, *tunturi* is used to refer to treeless plains at high altitudes in far north regions, as well.) They are round inselbergs rising from the otherwise flat surroundings. The mountains in Finnish Lapland reach heights of up to 400 and 800 metres, where the upper reaches are above the tree line. Those that do not reach the tree line, on the other hand, are mostly referred to as *vaara*. The mountains in Finnish Lapland form vestiges of the Karelides mountains, formed two billion years ago. The term *tunturi*, originally a word limited to Far-Northern dialects of Finnish and Karelian, is a loan from Sami, compare Proto-Sami **tuontër*, South Sami *doedtere*, Northern Sami *duottar*, Inari Sami *tuodâr* "uplands, mountains, tundra", Kildin Sami *tūndâr*, which means "uplands, treeless mountain tract" and is cognate with Finnish *tanner* "hard ground". From this Sami word, the word *tundra* is borrowed, as well, through the Russian language.

Förfjäll

The term förfjäll (literally "fore-fell") is used in Sweden to denote mountainous zones lower and less dissected than the fell proper. On the other hand, its more pronounced relief, its often higher amount of plateaux and its coherent valley systems distinguishes the förfjäll also from the undulating hilly terrain (bergkullsterräng) and the plains with residual hills (bergkullslätt). Generally the förfjäll do not surpass 1000 m.a.s.l. As a geomorphic unit the förfjäll extends across Sweden as a 650 km long and 40 to 80 km broad belt from Dalarna in the south to Norrbotten in the north.

Scandinavian Mountains

The Scandinavian Mountains or the Scandes is a mountain range that runs through the Scandinavian Peninsula. The Scandinavian Mountains are often erroneously thought to be equivalent to the Scandinavian Caledonides. The western sides of the mountains drop precipitously into the North Sea and Norwegian Sea, forming the fjords of Norway, whereas to the northeast they gradually curve towards Finland. To the north they form the border between Norway and Sweden, still reaching 2,000 m high (6,600 ft) at the Arctic Circle. The mountain range just touches northwesternmost Finland, but are scarcely more than hills at their northernmost extension at the North Cape (*Nordkapp*).

The mountains are not very high, but are very steep at places; Galdhøpiggen in southern Norway is the highest peak, at 2,469 metres (8,100 ft), Kebnekaise has the highest peak on the Swedish side, at 2,104 m (6,903 ft), whereas Halti is the highest peak in Finland, at 1,324 m (4,344 ft).

The combination of a northerly location and moisture from the North Atlantic Ocean has caused the formation of many ice fields and glaciers. Temperature drops with increasing altitude, and permafrost becomes common from about 1,500 meters above sea level on their western slope in Southern Norway and at about 1,200 meter AMSL on their eastern slope in Southern Norway near the border with Sweden. In Northern Norway, permafrost becomes common from about 800 to 900 meters AMSL on the western slope and some 200- to 300-meters-lower on the eastern slope.

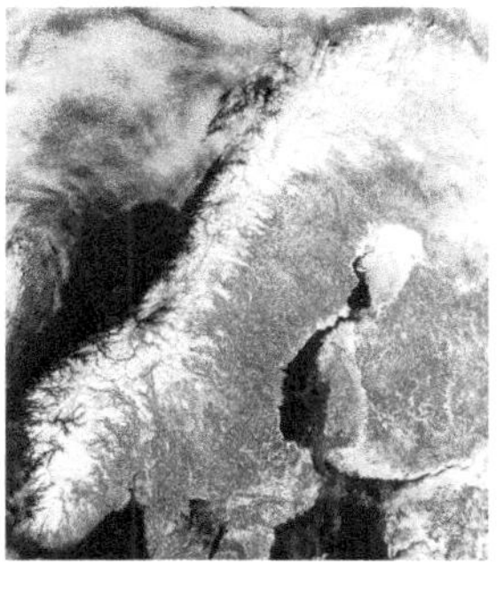

The Scandinavian Mountains show clearly in this satellite photo of the Scandinavian Peninsula from February 2003

The Scandinavian Montane Birch forest and grasslands terrestrial ecoregion is closely associated with the mountain range.

Names in Scandinavia

Its names in the Scandinavian languages are, in Swedish *Skanderna*, *Fjällen* ("the Fells") or *Kölen* (*Fjällen* being by far the most frequent name and virtually the only one in modern colloquial speech in Sweden) and in Norwegian *Kjølen*, with the three latter meaning **The Keel**.

Orography

The mountain chain's highest summits are mostly concentrated in an area (of mean altitude of over 1000 m) between Stavanger and Trondheim in southern Norway, with numerous peaks over 1300 m and some peaks over 2000 m. Around Trondheim Fjord peaks decrease in altitude to about 400–500 m rising again to heights in excess of 1900 m further north in Swedish Lapland and nearby areas of Norway. The southern part of the mountain range contains the highest mountain of Northern Europe, Galdhøpiggen at almost 2500 m. This part of the mountain chain is also broader and contains a series of plateaux and gently undulating surfaces that hosts scattered inselbergs. In south-western Norway the plateaux and gently undulating surfaces are strongly dissected by fjords and valleys. The mountain chain is present in Sweden from northern Dalarna northwards; south of this point the Scandinavian Mountains lie completely within Norway. Most of the Scandinavian Mountains lack alpine topography, and where it does have it it does relate to altitude.

To the east the Scandinavian Mountains proper bounds with mountains that are lower and less dissected and are known as the *förfjäll* (literally "fore-fell"). While lower than the Scandinavian Mountains proper, the förfjäll›s more pronounced relief, its often larger number of plateaux and its coherent valley systems distinguish it from so-called undulating hilly terrain (*bergkullsterräng*) and plains with residual hills (*bergkullslätt*) found further east. Generally the förfjäll do not surpass 1000 m.a.s.l. As a geomorphic unit the förfjäll extends across Sweden as a 650 km long and 40 to 80 km broad belt from Dalarna in the south to Norrbotten in the north.

Geology

Origin of the Mountains

The origin of today's mountain topography is debated by geologists. During the Paleozoic, a continental collision between Scandinavia (then part of Baltica) and Greenland (then part of Laurentia) produced a Himalayas-sized mountain range named the Caledonian Mountains roughly over the same area as the present-day Scandinavian Mountains. In Scandinavia the Caledonian Mountains begun a post-orogenic collapse in the Devonian implying tectonic extension and subsidence. Despite occurring in about the same area the ancient Caledonian Mountains and the modern Scandinavian Mountains are unrelated.

Geologically, the Scandivian Mountains are an elevated, passive continental margin similar to the mountains and plateaux found on the opposite side of the North Atlantic in Eastern Greenland or in Australia's Great Dividing Range. The Scandinavian Mountains attained its height by tectonic processes different from orogeny, chiefly in the Cenozoic. A two-stage model of uplift has been proposed for the Scandinavian Mountains in southern Norway. A first stage in the Mesozoic and a second stage starting from the Oligocene. The uplift of southern Norway has elevated the westernmost extension of the sub-Cambrian peneplain which forms part of what is known as the Paleic surface in Norway. In southern Norway the Scandinavian Mountains had their main uplift phase later (Neogene) than in northern Scandinavia which had its main phase of uplift in the Paleogene. The uplift in the north tilted the landsurface to the east, as reflected in the Muddus plains landscape.

Unlike orogenic mountains there is no geophysical model that explains elevated passive continental margins such as the Scandinavian Mountains. Various mechanisms of uplift have, however, been proposed over the years. Some have stressed the role of climate in inducing erosion that induces an isostatic compensation; fluvial and glacial erosion and incision during the Quaternary is thought to have contributed to the uplift of the mountain by forcing an isostatic response. The total amount of uplift produced by this mechanism could be as much as 500 m. Others geoscientists have implied diapirism in the asthenosphere as being the cause of uplift. One hypothesis claim that the early uplift of the Scandinavian Mountains could be indebted to changes in the density of the lithosphere and asthenosphere caused by the Iceland plume when Greenland and Scandinavia rifted apart about 53 million years ago.

Quaternary Geology

Many slopes and valleys have straight courses due to the fact that they follow tectonic fractures that are more prone to erosion. Another result of tectonics in the relief is that slopes corresponding to footwalls of normal faults tend to be straight.

There is evidence that the drainage divide between the Norwegian Sea and the south-east flowing rivers were once further west. Glacial erosion is thought to have contributed to the shift of the divide, which in some cases ought to have in excess of 50 km. Much of the Scandinavian Mountains has been sculpted by glacal erosion. The mountain chain is dotted with glacial cirques usually separated from each other by pre-glacialsurfaces. At some locations coalesced cirques form arêtes and pyramidal peaks. Glacial reshaping of valleys is more marked in the western part of the mountain chain where drowned glacier-shaped valleys constitute the fjords of Norway. In the eastern part of the mountain chain glacial reshaping of valleys is weaker. Many mountain tops contain blockfields which escaped glacial erosion either by having constituted nunataks in the glacial periods or by being protected from erosion under cold-based glacier ice. Karst systems, with their characteristic caves and sinkholes occur at various locations in the Scandinavian Mountains but are more common in the northern parts. Present-day karst systems might have long histories dating back to the Pleistocene or even earlier times.

Much of the mountain range is mantled by deposits of glacial origin including till blankets, moraines, drumlins and glacifluvial material in the form of outwash plains and eskers. Bare rock surfaces are more common in the western side of the mountain range. Although the ages of these deposits and landforms vary, most of them were formed in connection to the Weichselian glaciation and the subsequent deglaciation.

Highest Mountains

Norway

Of the 10 highest mountain peaks in Scandinavia (prominence greater than 30 m or 98 ft), six are situated in Oppland, Norway. The other four are situated in Sogn og Fjordane, Norway. There are 83 peaks equal to or taller than 2,200 m (7,218 ft) in Norway.

- 2,469 m (8,100 ft) Galdhøpiggen (Oppland)
- 2,465 m (8,087 ft) Glittertind (Oppland)
- 2,405 m (7,890 ft) Store Skagastølstind (Sogn og Fjordane)
- 2,387 m (7,831 ft) Store Styggedalstinden east (Sogn og Fjordane)
- 2,373 m (7,785 ft) Skarstind (Oppland)
- 2,369 m (7,772 ft) Vesle Galdhøpiggen (Oppland)
- 2,368 m (7,769 ft) Surtningssue (Oppland)
- 2,366 m (7,762 ft) Store Memurutinden (Oppland)
- 2,351 m (7,713 ft) Jervvasstind (Sogn og Fjordane)
- 2,348 m (7,703 ft) Sentraltind (Sogn og Fjordane)

Galdhøpiggen seen from west, Norway's highest mountain

Glittertind

Falketind in Jotunheimen

Landscape between Abisko National Park and Kebnekaise

Sweden

There are 12 peaks in Sweden that reach above 2,000 m high (6,600 ft), or 13 depending on how the peaks are defined. This list is defined according to UIAA. Eight of them are located in Sarek National Park and the neighbouring national park Stora Sjöfallet. The other four peaks are located in the further north region of Kebnekaise. All mountain names are in Sami but with the more common Swedish spelling of it.

- 2,104 m (6,903 ft) Kebnekaise (Lappland) - Note: Altitude includes the peak glacier. If melting continues, Kebnekaise Nordtoppen, just 500 meters away, might become the highest point.
- 2,097 m (6,880 ft) Kebnekaise Nordtoppen (Lappland) - the highest fixed point in Sweden.
- 2,089 m (6,854 ft) Sarektjåkkå Stortoppen (Lappland)
- 2,076 m (6,811 ft) Kaskasatjåkka (Lappland)
- 2,056 m (6,745 ft) Sarektjåkkå Nordtoppen (Lappland)
- 2,043 m (6,703 ft) Kaskasapakte (Lappland)
- 2,023 m (6,637 ft) Sarektjåkkå Sydtoppen (Lappland)
- 2,016 m (6,614 ft) Akka Stortoppen (Lappland)
- 2,010 m (6,594 ft) Akka Nordvästtoppen (Lappland)
- 2,010 m (6,594 ft) Sarektjåkkå Buchttoppen (Lappland)
- 2,005 m (6,578 ft) Pårtetjåkka (Lappland)
- 2,002 m (6,568 ft) Palkatjåkka (Lappland)

Other popular mountains for skiers, climbers and hikers in Sweden

- Sulitelma 1,860 m (Lappland)
- Helagsfjället 1,796 m (Härjedalen)
- Norra Storfjället 1,767 m (Lappland)
- Templet 1,728 m (Jämtland)
- Lillsylen 1,704 m (Jämtland)
- Åreskutan 1,420 m (Jämtland)
- Storvätteshågna 1,204 m (Dalarna)
- Molnet 1,191 m (Dalarna)

Finland

Landscape as seen from Meekonvaara (1019m) towards the highest fells

- 1,324 m (4,344 ft) Halti (Lappi/Lappland and Norwegian Troms)
- 1,317 m (4,321 ft) Ridnitsohkka (Lappi/Lappland)
- 1,280 m (4,200 ft) Kiedditsohkka (Lappi/Lappland)
- 1,240 m (4,068 ft) Kovddoskaisi (Lappi/Lappland)
- 1,239 m (4,065 ft) Ruvdnaoaivi (Lappi/Lappland)
- 1,180 m (3,871 ft) Loassonibba (Lappi/Lappland)
- 1,150 m (3,773 ft) Urtasvaara (Lappi/Lappland)
- 1,144 m (3,753 ft) Kahperusvaarat (Lappi/Lappland)
- 1,130 m (3,707 ft) Aldorassa (Lappi/Lappland)
- 1,100 m (3,608 ft) Kieddoaivi (Lappi/Lappland)

As of July 2016, to mark the centenary of Finland›s independence in 2017, the government of Norway is considering adjusting Norway›s border with Finland in the Halti area, giving Finland a 1,331 m peak (one of the two peaks of Halti), which would become Finland›s highest point.

The highest independent prominence in Finland is Taivaskero (807 m high) in Pallas-Yllästunturi National Park.

REFERENCES

1. Wainwright, A. (2003). "Coniston Old Man" in A Pictorial Guide to the Lakeland Fells, Book Four: The Southern Fells, p. 15. London: Francis Lincoln. ISBN 0-7112-2230-4
2. Bjordvand, Harald; Lindeman, Fredrik Otto (2007). Våre arveord. Novus. ISBN 978-82-7099-467-0

3. Falk, Hjalmar; Torp, Alf (2006). Etymologisk ordbog over det norske og det danske sprog. Bjørn Ringstrøms Antikvariat. ISBN 82-90520-16-6

CHAPTER 3

Moorland

Chapter Contents

Heather moorland is rare on a worldwide scale – there is probably less heather moorland in the world than tropical rainforest. Around 70 percent of the world's heather moorland is in the UK and the largest continuous expanse of moorland in England and Wales is here in the North York Moors – a sheep could wander from Egton to Bilsdale without leaving the moor. Moorland covers a third of the North York Moors National Park and most of the higher ground is covered in heather.

INTRODUCTION

Moorland or moor is a type of habitat found in upland areas in temperate grasslands, savannas, and shrublands and montane grasslands and shrublands biomes, characterised by low-growing vegetation on acidic soils. Moorland nowadays generally means uncultivated hill land (such as Dartmoor in South West England), but includes low-lying wetlands (such as Sedgemoor, also South West England). It is closely related to heath although experts disagree on precisely what distinguishes the types of vegetation. Generally, moor refers to highland, high rainfall zones, whereas heath refers to lowland zones which are more likely to be the result of human activity.

Moorland habitats are most extensive in the neotropics and tropical Africa but also occur in northern and western Europe, Northern Australia, North America, Central Asia, South America and the Indian subcontinent. Most of the world's moorlands are very diverse ecosystems. In the extensive moorlands of the tropics biodiversity can be extremely high. Moorland also bears a relationship to tundra (where the subsoil is permafrost or permanently frozen soil), appearing as the tundra retreats and inhabiting the area between the permafrost and the natural tree zone. The boundary between tundra and moorland constantly shifts with climate change.

Extensive moorland in the Desert of Wales

HEATHER MOORLAND

Heather moorland on the **North York Moors** mainly consisting of *Calluna vulgaris*

Heathland and moorland are the most extensive areas of semi-natural vegetation in the British Isles. The eastern British moorlands are similar to heaths but are differentiated by having a covering of peat. On western moors the peat layer may be several metres thick. Scottish "muirs" are generally heather moors, but also have extensive covering of grass, cotton-grass, mosses, bracken and under-shrubs such as crowberry, with the wetter moorland having sphagnum moss merging into bog-land.

There is uncertainty about how many moors were created by human activity. Rackham writes that pollen analysis shows that some moorland, such as in the islands and extreme north of Scotland, are clearly natural, never having had trees, whereas much of the Pennine moorland area was forested in Mesolithic times. How much the deforestation was caused by climatic changes and how much by human activity is uncertain.

Ecology

A variety of distinct habitat types are found in different world regions of moorland. The wildlife and vegetation forms often lead to high endemism because of the severe soil and microclimate characteristics. For example, in England›s Exmoor is found the rare horse breed the Exmoor Pony, which has adapted to the harsh conditions of that environment.

In Europe, the associated fauna consists of bird species such as red grouse, hen harrier, merlin, golden plover, curlew, skylark, meadow pipit, whinchat, ring ouzel, and twite. Other species dominate in moorlands elsewhere. Reptiles are few due to the cooler conditions. In Europe, only the common viper is frequent, though in other regions moorlands are commonly home to dozens of reptile species. Amphibians such as frogs are well represented in moorlands. When moorland is overgrazed, woody vegetation is often lost, being replaced by coarse, unpalatable grasses and bracken, with a greatly reduced fauna.

Moorland of **Kilimanjaro**

Some hill sheep breeds, such as Scottish Blackface and the Lonk, thrive on the austere conditions of heather moors.

Management

Burning of moorland has been practiced for a number of reasons, for example when grazing is insufficient to control growth. This is recorded in Britain in the fourteenth century. Uncontrolled burning frequently caused (and causes) problems, and was forbidden by statute in 1607. With the rise of sheep and grouse management in the nineteenth century it again became common practice. Heather is burnt at about 10 or 12 years old when it will regenerate easily. Left longer, the woodier stems will burn more aggressively and will hinder regrowth. Burning of moorland vegetation needs to be very carefully controlled as the peat itself can catch fire, and this can be difficult if not impossible to extinguish. In addition, uncontrolled burning of heather can promote alternative bracken and rough grass growth which ultimately produces poorer grazing. As a result, burning is now a controversial practice; Rackham calls it «second-best land management».

Mechanical cutting of the heather has been used in Europe, but it is important for the material to be removed to avoid smothering regrowth. If heather and other vegetation is left for too long, a large volume of dry and combustible material builds up. This may result in a wildfire burning out a large area, although it has been found that heather seeds germinate better if subject to the brief heat of controlled burning.

Moorland in Literature

Dartmoor in winter covered in snow. Several **tors** top the sparsely vegetated hills.

The development of a sensitivity to nature and one's physical surroundings grew with the rise of interest in landscape painting, and particularly the works of artists that favoured wide and deep prospects, and rugged scenery. To the English Romantic imagination, moorlands fitted this image perfectly, enhancing the emotional impact of the story by placing it within a heightened and evocative landscape. Moorland forms the setting of various works of late Romantic English literature, ranging from the Yorkshire moorland in Emily Brontës Wuthering Heights and The Secret Garden by Frances Hodgson Burnett to Dartmoor in Arthur Conan Doyle›s Holmesian mystery The Hound of the Baskervilles.

Enid Blyton's Famous Five series featured the young protagonists adventuring across various moorlands where they confronted criminals or other individuals of interest. Such a setting enhanced the plot as the drama unfolded away from the functioning world where the children could solve their own problems and face greater danger. Moorland in the Forest of Bowland in Lancashire is the setting for Walter Bennett›s The Pendle Witches, the true story of some of England's most infamous witch trials. In Erin Hunter's Warriors series, one of the four Clans, WindClan, lives in the moorland alone.

Michael Jecks, author of Knights Templar Mysteries, sets his books in and around Dartmoor, England.

NOTABLE MOORLANDS

The United Kingdom is home to an estimated 10–15% of the world's moors. Notable areas of upland moorland in Britain include the Dark Peak, the Forest of Bowland, the Lake District, the Pennines, Mid Wales, the Southern Uplands of Scotland, the Scottish Highlands, and a few very small pockets in western Herefordshire.

Oswaldtwistle Moor, part of the West Pennine Moors, in Lancashire, UK

Bleaklow

Bleaklow is a high, largely peat-covered, gritstone moorland, just north of Kinder Scout, across the Snake Pass (A57), in the Derbyshire High Peak near the town of Glossop. Much of it is nearly 2,000 feet (610 m) above sea level and the shallow bowl of Swains Greave on its eastern side is the source of the River Derwent.

Bleaklow Head (633 m), marked by a huge cairn of stones, is the high point at the western side of the moor, is a Hewitt and is crossed by the Pennine Way. It is one of three summits on this plateau above 2,000 feet, the others being Bleaklow Stones, some 1.9 miles (3 km) to the east along an indefinite ridge, and Higher Shelf Stones, 0.9 miles (1.5 km) south of Bleaklow Head. At 633 metres (2,077 feet), Bleaklow is the second-highest point in Derbyshire and the area includes the most easterly point in the British Isles over 2,000 feet, near Bleaklow Stones.

Description

Much of the main plateau of Bleaklow is a boggy peat moorland, seamed by 'groughs' (pronounced 'gruffs', water-eroded channels in the peat), and lacking strong changes in elevation – in poor conditions its traverse is probably the most navigationally challenging in the Peak District.

Conservation

Bleaklow is part of the National Trust's High Peak Estate. There has been considerable investment of resources in recent years to block many of the eroded peat gulleys as part of major schemes to re-wet and restore healthy Sphagnum moss communities which are essential for peat formation, carbon-capture, and reduction in dissolved carbon which contaminates water supplies. Much of this work has been coordinated by the Moors for the Future project, funded by EU LIFE+ programme between 2010-2015. This involved laying 52 kilometres of geotextiles to stabilise eroded peat, creating 4,000 mini dams to retain water, introducing 150,000 moorland plants and spreading 807 million Sphagnum fragments across the whole Bleaklow Project site.

Views

The summit affords views across Manchester, Lancashire and Cheshire to the west and towards the Hope Valley, Holme Moss, Emley Moor and Yorkshire to the east. In exceptional weather conditions it is possible to see Snowdonia in North Wales.

Aircraft Crash

On 3 November 1948, USAF Boeing RB-29A Superfortress 44-61999, of the 16th Photographic Reconnaissance Squadron, 91st Reconnaissance Group, 311th Air Division, Strategic Air Command; crashed at Higher Shelf Stones, Bleaklow, whilst en route from Scampton to Burtonwood. All 13 crew members were killed. A large amount of wreckage is still visible, as a memorial to the crash. Also, a proper memorial was erected at the site in 1988. There is public access to the area.

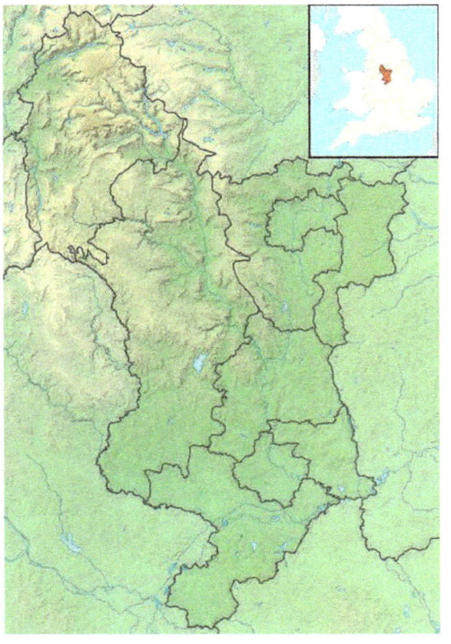

Map showing the location of Bleaklow within Derbyshire

Gallery

Memorial plaque by the wreckage

Trig point on Shelf Stones

Descending eroded peat gully on east flank of Bleaklow.

Dark Peak

The Dark Peak is the higher, wilder northern part of the Peak District in England, mainly in Derbyshire and South Yorkshire.

It gets its name because (in contrast to the White Peak), the underlying limestone is covered by a cap of Millstone Grit which means that in winter the soil is almost always saturated with water. The land is thus largely uninhabited moorland plateaux where almost any depression is filled with sphagnum bogs and black peat. The High Peak is an alternative name for the Dark Peak, but High Peak is also the name of an administrative district of Derbyshire which includes part of the White Peak.

The areas of Millstone Grit form an ‹inverted horseshoe' around the lower uncapped limestone areas of the White Peak. Hence the Dark Peak is said to cover the higher Derbyshire Peaks north of the Hope Valley, as well as the Western Moors in the area of Buxton, and the Eastern Moors stretching south towards Matlock. The Dark Peak is one of 159 National Character Areas defined by Natural England; as defined by Natural England, the Dark Peak NCA covers 86,604 hectares (334 sq mi) and includes the northern block of hills approximately bounded by Marsden, Stocksbridge, Hathersage and Chapel-en-le-Frith, plus the eastern moors south to Matlock, but excludes the western moors south of Chapel (which it places in NCA 53, the South West Peak) and the area around Glossop (in NCA 54, Manchester Pennine Fringe).

An area of 31,852 hectares (123 sq mi) is designated as the Dark Peak Site of Special Scientific Interest (SSSI), which excludes the separately designated Eastern Moors. The SSSI extends over the borders into Greater Manchester and West Yorkshire. A large part of the SSSI is included in the South Pennine Moors Special Area of Conservation.

The summit of Bleaklow, second highest hill in the Dark Peak

Principal upland areas within the Dark Peak include Kinder Scout, Bleaklow (both of which rise to over 2000 ft, and lie wholly within Derbyshire), and Black Hill.

Over the years, a number of military aircraft have crashed on the Dark Peak, generally due to atrocious weather conditions. Due to the bleakness and emptiness of the high moorlands, their wrecks tend to remain untouched, and are often stumbled across by hikers in these remote places. There have been reports of ghost planes in the area for years. Many of the aircraft witnessed are wartime machines such as the Wellington Bomber, B-29 Bomber, Dakota and Lancaster. The reports of eyewitnesses are usually of a low flying, propeller-driven plane, which appears to be in difficulty before seemingly crashing into the moors. Several people who recovered items from the crash site were allegedly then visited by ghosts.

Bodmin Moor

Bodmin Moor (Cornish: Goon Brenn) is a granite moorland in northeastern Cornwall, England. It is 208 square kilometres (80 sq mi) in size, and dates from the Carboniferous period of geological history. It includes Brown Willy, the highest point in Cornwall, and Rough Tor, a slightly lower peak. Many of Cornwall's rivers have their sources here. It has been inhabited since at least the Neolithic era, when primitive farmers started clearing trees and farming the land. They left their megalithic monuments, hut circles and cairns, and the Bronze Age culture that followed left further cairns, and more stone circles and stone rows. By medieval and modern times, nearly all the forest was gone and livestock rearing predominated.

The name Bodmin Moor is relatively recent, an Ordnance Survey invention of 1813. The upland area was formerly known as Fowey Moor after the River Fowey, which rises within it.

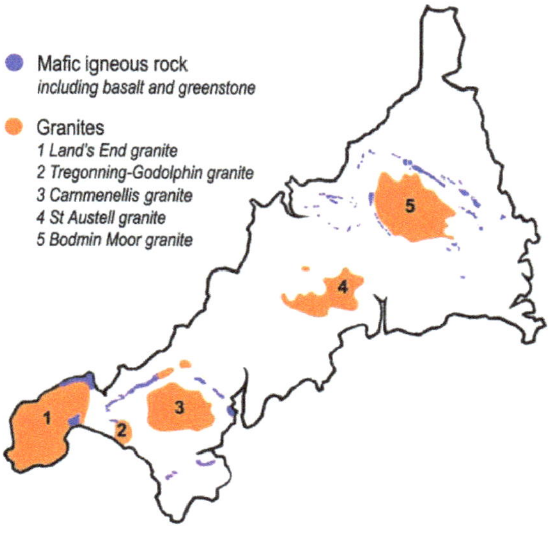

Geological sketch showing Bodmin Moor (5) in relation to Cornwall's granite intrusions

Rough Tor

Geography

Bodmin Moor is one of five granite plutons in Cornwall that make up part of the Cornubian batholith .

Dramatic granite tors rise from the rolling moorland: the best known are Brown Willy, the highest point in Cornwall at 417 m (1,368 ft), and Rough Tor at 400 m (1,300 ft). To the south-east Kilmar Tor and Caradon Hill are the most prominent hills. Considerable areas of the moor are poorly drained and form marshes (in hot summers these can dry out). The rest of the moor is mostly rough pasture or overgrown with heather and other low vegetation.

The moor contains about 500 holdings with around 10,000 beef cows, 55,000 breeding ewes and 1,000 horses and ponies. Most of the moor is a Site of Special Scientific Interest (SSSI), Bodmin Moor, North, and has been officially designated an Area of Outstanding Natural Beauty (AONB), as part of Cornwall AONB. Almost a third of Cornwall has AONB designation, with the same status and protection as a National Park. The moor has been identified by BirdLife International as an Important Bird Area (IBA) because it supports about 260 breeding pairs of European stonechats as well as a wintering population of 10,000 Eurasian golden plovers. The moor has also been recognised as a separate natural region and designated as national character area 153 by Natural England.

Rivers and Inland Waters

Siblyback Lake

The **De Lank River** at **Garrow Tor**

Bodmin Moor is the source of several of Cornwall's rivers: they are mentioned here anti-clockwise from the south.

The River Fowey rises at a height of 290 m (950 ft) and flows through Lostwithiel and into the Fowey estuary.

The River Tiddy rises near Pensilva and flows southeast to its confluence with the River Lynher (the Lynher flows generally south-east until it joins the Hamoaze near Plymouth). The River Inny rises near Davidstow and flows southeast to its confluence with the River Tamar.

The River Camel rises on Hendraburnick Down and flows for approximately 40 km (25 mi) before joining the sea at Padstow. The River Camel and its tributary the De Lank River are an important habitat for the otter and both have been proposed as Special Areas of Conservation (SAC) The De Lank River rises near Roughtor and flows along an irregular course before joining the Camel south of Wenford.

The River Warleggan rises near Temple and flows south to join the Fowey.

On the southern slopes of the moor lies Dozmary Pool. It is Cornwall's only natural inland lake and is glacial in origin. In the 20th century three reservoirs have been constructed on the moor; these are Colliford Lake, Siblyback Lake and Crowdy reservoirs which supply water for a large part of the county›s population. Various species of waterfowl are resident around these waters.

Rivers south of Bodmin Moor

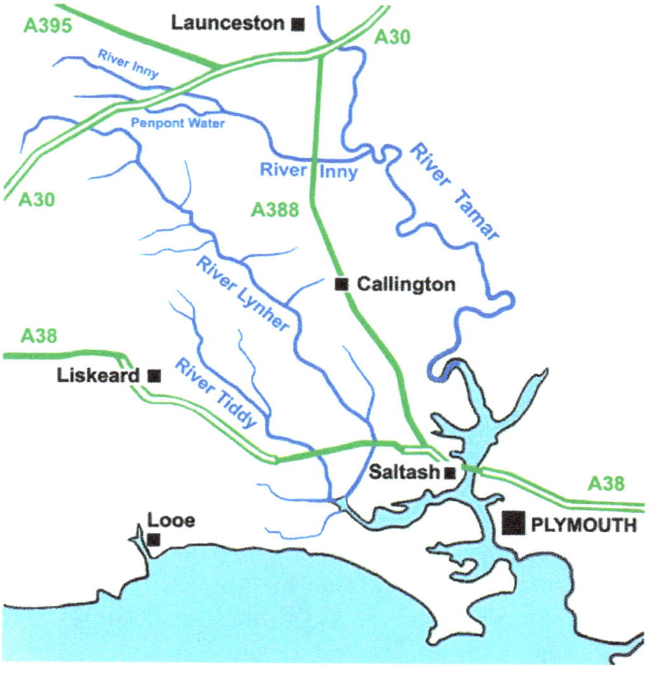

Rivers southeast of Bodmin moor

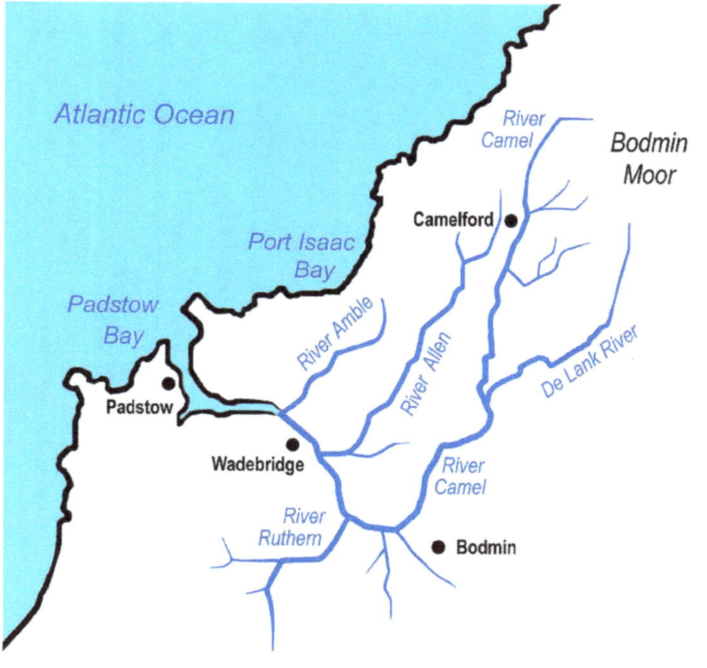

Rivers northwest of Bodmin Moor

Parishes

Church in **St Neot**

The parishes on the moor are as follows:

• Advent	• North Hill
• Altarnun	• St Breward
• Blisland	• St Cleer
• Bolventor	• St Clether
• Camelford	• St Ive
• Davidstow	• St Neot
• Lewannick	• Warleggan
• Linkinhorne	

Prehistoric Times

King Arthur's Hall

Kilmar Tor

10,000 years ago, in the Mesolithic period, hunter-gatherers wandered the area when it was wooded. There are several documented cases of flint scatters being discovered by archaeologists, indicating that these hunter gatherers practised flint knapping in the region.

During the Neolithic era, from about 4,500 to 2,300 BC, people began clearing trees and farming the land. It was also in this era that the production of various megalithic monuments began, predominantly long cairns (three of which have currently been identified, at Louden, Catshole and Bearah) and stone circles (sixteen of which have been identified). It was also likely that the naturally forming tors were also viewed in a similar manner to the manmade ceremonial sites.

In the following Bronze Age, the creation of monuments increased dramatically, with the production of over 300 further cairns, and more stone circles and stone rows. More than 200 Bronze Age settlements with enclosures and field patterns have been recorded. and many prehistoric stone barrows and circles lie scattered across the moor. In a programme shown in 2007 Channel 4›s Time Team investigated a 500-metre cairn and the site of a Bronze Age village on the slopes of Rough Tor.

King Arthur's Hall thought to be a late Neolithic or early Bronze Age ceremonial site can be found to the east of St Breward on the moor.

Medieval and Modern Times

Hawk's Tor, west of **North Hill**

Where practicable areas of the moor were used for pasture by herdsmen from the parishes surrounding the moor. Granite boulders were also taken from the moor and used for stone posts and to a certain extent for building (such material is known as moorstone). Granite quarrying only became reasonably productive when gunpowder became available.

The moor gave its name (Foweymore) to one of the medieval districts called **stannaries** which administered tin mining: the boundaries of these were never defined precisely. Until the establishment of a turnpike road through the moor (the present **A30**) in the 1770s the size of the moorland area made travel within Cornwall very difficult.

Its Cornish name, Goen Bren, is first recorded in the 12th century.

English Heritage monographs "Bodmin Moor: An Archaeological Survey" Volume 1 and Volume 2 covering the post-medieval and modern landscape are publicly available through the Archaeology Data Service.

Monuments and Ruins

Roughtor was the site of a medieval chapel of St Michael and is now designated as a memorial to the 43rd Wessex Division of the British Army. In 1844 on Bodmin Moor the body of 18-year-old Charlotte Dymond was discovered. Local labourer Matthew Weeks was accused of the murder and at noon on 12 August 1844 he was led from Bodmin Gaol and hanged. The murder site now has a monument erected from public money and the grave is at Davidstow churchyard.

Legends and Traditions

Dozmary Pool is identified by some people with the lake in which, according to Arthurian legend, Sir Bedivere threw Excalibur to The Lady of the Lake. Another legend relating to the pool concerns Jan Tregeagle.

The Beast of Bodmin has been reported many times but never identified with certainty.

Cornwall

Cornwall is a ceremonial county and unitary authority area of England within the United Kingdom. It is bordered to the north and west by the Celtic Sea, to the south by the English Channel, and to the east by the county of Devon, over the River Tamar. Cornwall has a population of 536,000 and covers an area of 3,563 km^2(1,376 sq mi). The administrative centre, and only city in Cornwall, is Truro, although the town of Falmouth has the largest population for a civil parish and the conurbation of Camborne, Pool and Redruth has the highest total population.

Cornwall forms the westernmost part of the south-west peninsula of the island of Great Britain, and a large part of the Cornubian batholith is within Cornwall. This area was first inhabited in the Palaeolithic and Mesolithic periods. It continued to be occupied by Neolithic and then Bronze Age peoples, and later (in the Iron Age) by Brythons with distinctive cultural relations to neighbouring Wales and Brittany. There is little evidence that Roman rule was effective west of Exeter and few Roman remains have been found. Cornwall was the home of a division of the Dumnonii tribe – whose tribal centre was in the modern county of Devon – known as the Cornovii, separated from the Brythons of Wales after the Battle of Deorham, often coming into conflict with the expanding kingdom of Wessex before King Athelstan in AD 936 set the boundary between English and Cornish at the high water mark of the eastern bank of the River Tamar. From the early Middle Ages, British language and culture was apparently shared by Brythons trading across both sides of the Channel, evidenced by the corresponding high medieval

Breton kingdoms of Domnonée and Cornouaille and the Celtic Christianity common to both territories.

Historically tin mining was important in the Cornish economy, becoming increasingly significant during the High Middle Ages and expanding greatly during the 19th century when rich copper mines were also in production. In the mid-19th century, however, the tin and copper mines entered a period of decline. Subsequently, china clay extraction became more important and metal mining had virtually ended by the 1990s. Traditionally, fishing (particularly of pilchards) and agriculture (notably dairy products and vegetables) were the other important sectors of the economy. Railways led to a growth of tourism in the 20th century; however, Cornwall's economy struggled after the decline of the mining and fishing industries. The area is noted for its wild moorland landscapes, its long and varied coastline, its attractive villages, its many place-names derived from the Cornish language, and its very mild climate. Extensive stretches of Cornwall's coastline, and Bodmin Moor, are protected as an Area of Outstanding Natural Beauty.

Cornwall is the homeland of the Cornish people and is recognised as one of the Celtic nations, retaining a distinct cultural identity that reflects its history. Some people question the present constitutional status of Cornwall, and a nationalist movement seeks greater autonomy within the United Kingdom in the form of a devolved legislative Cornish Assembly. On 24 April 2014 it was announced that Cornish people will be granted minority status under the European Framework Convention for the Protection of National Minorities.

Toponymy

The name Cornwall derives from the combination of two separate terms from different languages. The Corn- part comes from the hypothesised original tribal name of the Celtic people who had lived here since the Iron Age, the Cornovii. The second element -wall derives from the Old English w(e)alh, meaning a "foreigner" or "Welshman". The name first appears in the Anglo-Saxon Chronicle in 891 as On Corn walum. In the Domesday Book it was referred to as Cornualia and in c. 1198 as Cornwal.

A latinisation of the name as Cornubia first appears in a mid-9th-century deed purporting to be a copy of one dating from c. 705. Another variation, with Wales reinterpreted as Gallia, thus: Cornugallia, is first attested in 1086. Finally, the Cornish language form of the name, Kernow, which first appears around 1400, derives directly from the original Cornowii. which is postulated from a single mention in the Ravenna Cosmography of around 700 (but based on earlier sources) of Purocoronavis. This is considered to be a corruption of Durocornovium, 'a fort or walled settlement of the Cornovii'. Its location is unidentified, but Tintagel or Carn Brea have been suggested.

In pre-Roman times, Cornwall was part of the kingdom of Dumnonia, and was later known to the Anglo-Saxons as "West Wales", to distinguish it from "North Wales" (modern-day Wales).

Cornwall is one of only a few places in Britain – London, Edinburgh, and Dover being other examples – to have a corresponding name in the French language: *Cornouailles* (French pronunciation:

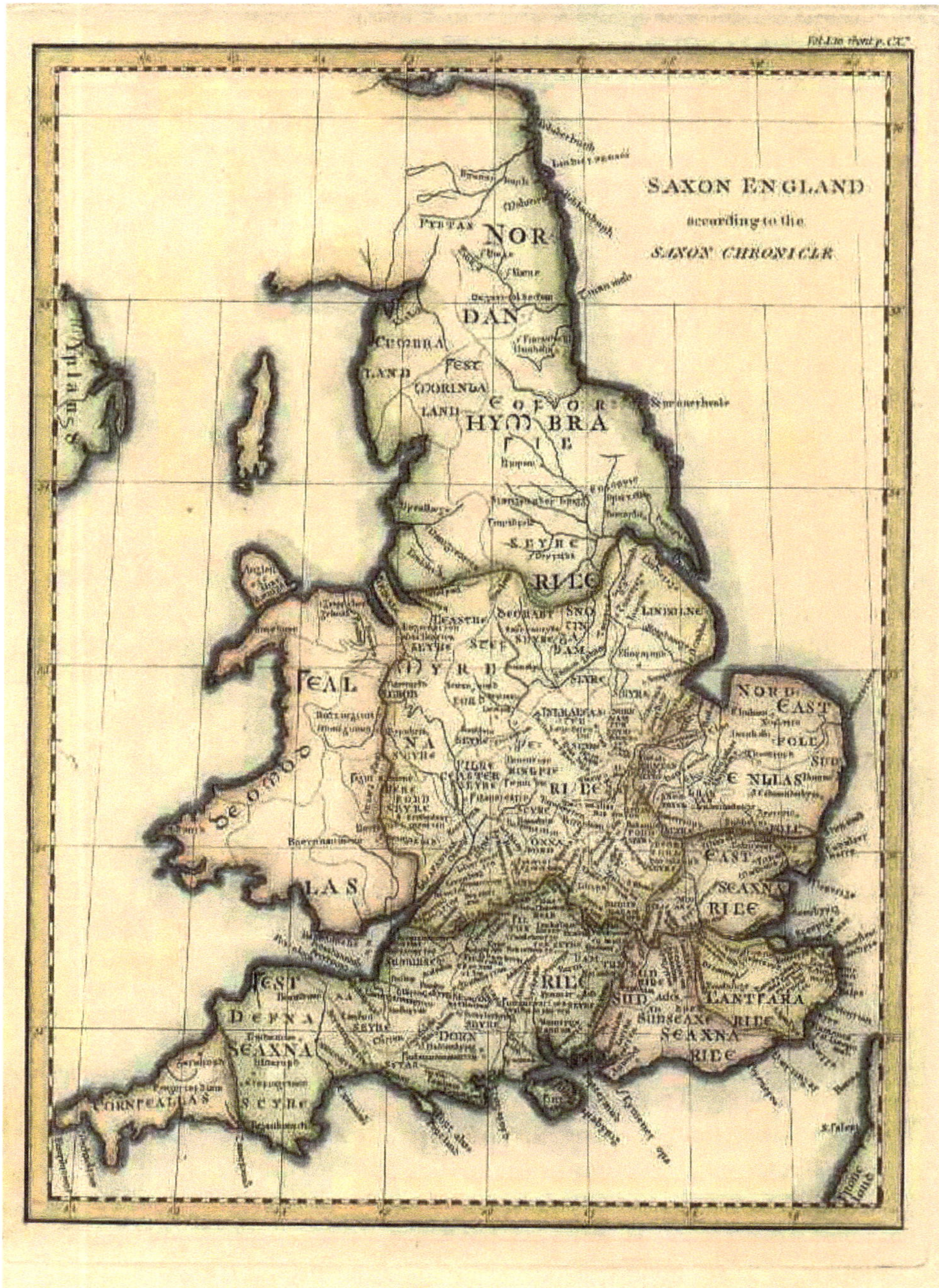

"Cornweallas" shown on an early 19th-century map of "Saxon England" (and Wales) based on the ***Anglo-Saxon Chronicle***.

Mên-an-Tol.

Prehistory, Roman and Post-Roman Periods

The present human history of Cornwall begins with the reoccupation of Britain after the last Ice Age. The area now known as Cornwall was first inhabited in the Palaeolithic and Mesolithic periods. It continued to be occupied by Neolithic and then Bronze Age people. According to John T. Koch and others, Cornwall in the Late Bronze Age was part of a maritime trading-networked culture called the Atlantic Bronze Age, in modern-day Ireland, England, France, Spain and Portugal. During the British Iron Age, Cornwall, like all of Britain south of the Firth of Forth, was inhabited by a Celtic people known as the Britons with distinctive cultural relations to neighbouring Wales and Brittany. The Common Brittonic spoken at the time eventually developed into several distinct tongues, including Cornish.

The first account of Cornwall comes from the 1st century BC Sicilian Greek historian Diodorus Siculus, supposedly quoting or paraphrasing the 4th-century BCE geographer Pytheas, who had sailed to Britain:

The inhabitants of that part of Britain called Belerion (or Land's End) from their intercourse with foreign merchants, are civilised in their manner of life. They prepare the tin, working very carefully the earth in which it is produced ... Here then the merchants buy the tin from the natives and carry it over to Gaul, and after travelling overland for about thirty days, they finally bring their loads on horses to the mouth of the Rhône.

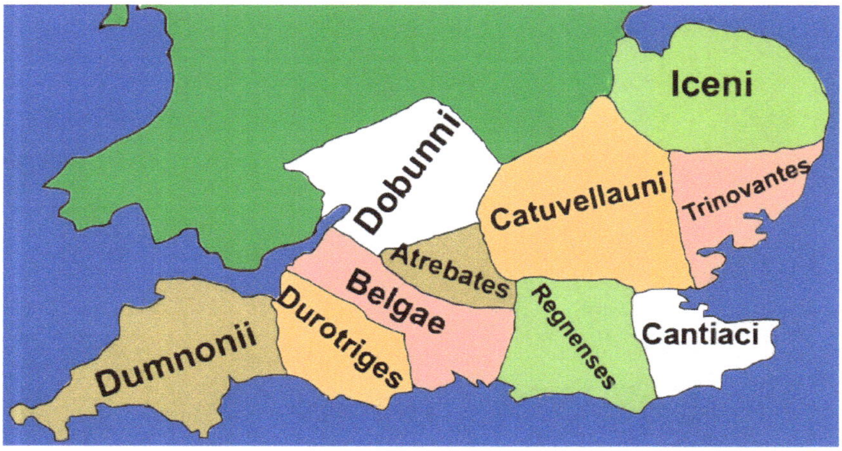

Celtic tribes of Southern Britain.

The identity of these merchants is unknown. It has been theorised that they were Phoenicians, but there is no evidence for this. Professor Timothy Champion, discussing Diodorus Siculus's comments on the tin trade, states that "Diodorus never actually says that the Phoenicians sailed to Cornwall. In fact, he says quite the opposite: the production of Cornish tin was in the hands of the natives of Cornwall, and its transport to the Mediterranean was organised by local merchants, by sea and then over land through France, well outside Phoenician control." (For further discussion of tin mining see the section on the economy below.)

There is little evidence that Roman rule was effective west of Exeter in Devon and few Roman remains have been found. However, after 410, Cornwall appears to have reverted to rule by Romano-Celtic chieftains of the Cornovii tribe as part of Dumnonia including one Marcus Cunomorus with at least one significant power base at Tintagel. ‹King' Mark of Cornwall is a semi-historical figure known from Welsh literature, the Matter of Britain, and in particular, the later Norman-Breton medieval romance of Tristan and Yseult where he is regarded as a close kinsman of King Arthur; himself usually considered to be born of the Cornish people in folklore traditions derived from Geoffrey of Monmouth's Historia Regum Britanniae. Archaeology supports ecclesiastical, literary and legendary evidence for some relative economic stability and close cultural ties between the sub-Roman Westcountry, South Wales, Brittany and Ireland through the fifth and sixth centuries.

Conflict with Wessex

The Battle of Deorham in 577 saw the separation of Dumnonia (and therefore Cornwall) from Wales, following which the Dumnonii often came into conflict with the expanding English kingdom of Wessex. The Annales Cambriae report that in 722 AD the Britons of Cornwall won a battle at «Hehil». It seems likely that the enemy the Cornish fought was a West Saxon force, as evidenced by the naming of King Ine of Wessex and his kinsman Nonna in reference to an earlier Battle of Lining in 710.

The Anglo-Saxon Chronicle stated in 815 (adjusted date) "and in this year king Ecgbryht raided in Cornwall from east to west." and thenceforth apparently held it as a ducatus or dukedom annexed to his regnum or kingdom of Wessex, but not wholly incorporated with it. The Anglo-Saxon Chronicle states that in 825 (adjusted date) a battle took place between the Wealas (Cornish) and the Defnas (men of Devon) at Gafulforda. In the same year Ecgbert, as a later document expresses it, "disposed of their territory as it seemed fit to him, giving a tenth part of it to God." In other words, he incorporated Cornwall ecclesiastically with the West Saxon diocese of Sherborne, and endowed Eahlstan, his fighting bishop, who took part in the campaign, with an extensive Cornish estate consisting of Callington and Lawhitton, both in the Tamar valley, and Pawton near Padstow.

In 838, the Cornish and their Danish allies were defeated by Egbert in the Battle of Hingston Down at Hengestesdune (probably Hingston Down in Cornwall). In 875, the last recorded king of Cornwall, Dumgarth, is said to have drowned. Around the 880s, Anglo-Saxons from Wessex had established modest land holdings in the eastern part of Cornwall; notably Alfred the Great who had acquired a few estates. William of Malmesbury, writing around 1120, says that King Athelstan of England (924–939) fixed the boundary between English and Cornish people at the east bank of the River Tamar.

Breton–Norman Period

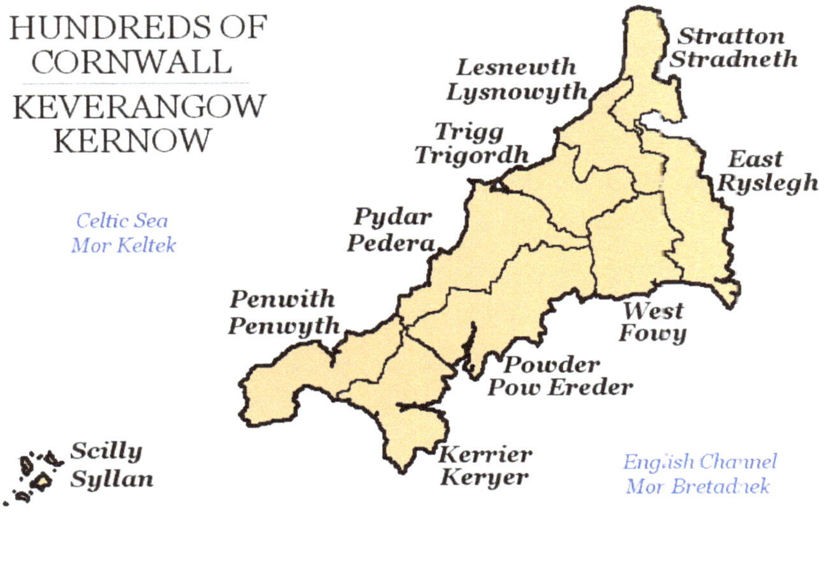

The ancient **Hundreds of Cornwall**

One interpretation of the Domesday Book is that by this time the native Cornish landowning class had been almost completely dispossessed and replaced by English landowners, particularly Harold Godwinson himself. However, the Bodmin manumissions show that two leading Cornish figures nominally had Saxon names, but these were both glossed with native Cornish names. Naming evidence cited by medievalist Edith Ditmas suggests that many post-Conquest landowners in Cornwall were Breton allies of the

Normans and further proposed this period for the early composition of the Tristan and Iseult cycle by poets such as Beroul from a pre-existing shared Brittonic oral tradition.

Soon after the Norman conquest most of the land was transferred to the new Breton–Norman aristocracy, with the lion's share going to Robert, Count of Mortain, half-brother of King William and the largest landholder in England after the king with his stronghold at Trematon Castle near the mouth of the Tamar. Cornwall and Devon west of Dartmoor showed a very different type of settlement pattern from that of Saxon Wessex and places continued, even after 1066, to be named in the Celtic Cornish tradition with Saxon architecture being uncommon.

Later Medieval Administration and Society

Subsequently, however, Norman absentee landlords became replaced by a new Cornu-Norman elite including scholars such as Richard Rufus of Cornwall. These families eventually became the new ruling class of Cornwall (typically speaking Norman French, Breton-Cornish, Latin and eventually English), many becoming involved in the operation of the Stannary Parliament system, Earldom and eventually the Duchy. The Cornish language continued to be spoken and it acquired a number of characteristics establishing its identity as a separate language from Breton.

Cornish piracy was active during the Elizabethan era on the west coast of Britain.

Christianity in Cornwall

Many place names in Cornwall are associated with Christian missionaries described as coming from Ireland and Wales in the 5th century AD and usually called saints (See List of Cornish saints). The historicity of some of these missionaries is problematic. The patron saint of Wendron Parish Church, "Saint Wendrona" is another example. and it has been pointed out by Canon Doble that it was customary in the Middle Ages to ascribe such geographical origins to saints. Some of these saints are not included in the early lists of saints.

Saint Piran, after whom Perranporth is named, is generally regarded as the patron saint of Cornwall. However, in early Norman times it is likely that Saint Michael the Archangel was recognised as the patron saint and is still recognised by the Anglican Church as the Protector of Cornwall. The title has also been claimed for Saint Petroc who was patron of the Cornish diocese prior to the Normans.

Celtic and Anglo-Saxon Times

St German's Priory Church (Norman)

Dupath Well, one of Cornwall's many holy wells dating from c.1510

The Church of St Petroc at Bodmin (late 15th century)

The church in Cornwall until the time of Athelstan of Wessex observed more or less orthodox practices, being completely separate from the Anglo-Saxon church until then (and perhaps later). The See of Cornwall continued until much later: Bishop Conan apparently in place previously, but (re-?) consecrated in 931 AD by Athelstan. However, it is unclear whether he was the sole Bishop for Cornwall or the leading Bishop in the area. The situation in Cornwall may have been somewhat similar to Wales where each major religious house corresponded to a cantref (this has the same meaning as Cornish keverang) both being under the supervision of a Bishop. However, if this was so the status of keverangow before the time of King Athelstan is not recorded. However, it can be inferred from the districts included at this period that the minimum number would be three: Triggshire; Wivelshire; and the remaining area. Penwith, Kerrier, Pydar and Powder meet at a central point (Scorrier) which some have believed indicates a fourfold division imposed by Athelstan on a sub-kingdom.

Middle Ages

The whole of Cornwall was in this period in the Archdeaconry of Cornwall within the Diocese of Exeter. From 1267 the archdeacons had a house at Glasney near Penryn. Their duties were to visit and inspect each parish annually and to execute the bishop›s orders. Archdeacon Roland is recorded in the Domesday Book of 1086 as having land holdings in Cornwall but he was not Archdeacon of Cornwall, just an archdeacon in the Diocese of Exeter. In the episcopate of William Warelwast (1107–37) the first Archdeacon of Cornwall was appointed (possibly Hugo de Auco). Most of the parish churches in Cornwall in Norman times were not in the larger settlements, and the medieval towns which developed thereafter usually had only a chapel of ease with the right of burial remaining at the ancient parish church. Over a hundred holy wells exist in Cornwall, each associated with a particular saint, though not always the same one as the dedication of the church.

Various kinds of religious houses existed in mediaeval Cornwall though none of them were nunneries; the benefices of the parishes were in many cases appropriated to religious houses within Cornwall or elsewhere in England or France.

From the Reformation to the Victorian Period

In the 16th century there was some violent resistance to the replacement of Catholicism with Protestantism in the Prayer Book Rebellion. In 1548 the college at Glasney, a centre of learning and study established by the Bishop of Exeter, had been closed and looted (many manuscripts and documents were destroyed) which aroused resentment among the Cornish. They, among other things, objected to the English language Book of Common Prayer, protesting that the English language was still unknown to many at the time. The Prayer Book Rebellion was a cultural and social disaster for Cornwall; the reprisals taken by the forces of the Crown have been estimated to account for 10–11% of the civilian population of Cornwall. Culturally speaking, it saw the beginning of the slow decline of the Cornish language.

From that time Christianity in Cornwall was in the main within the Church of England and subject to the national events which affected it in the next century and a half. Roman Catholicism never became extinct, though openly practised by very few; there were some converts to Puritanism, Anabaptism and Quakerism in certain areas though they suffered intermittent persecution which more or less came to an end in the reign of William and Mary. During the 18th century Cornish Anglicanism was very much in the same state as Anglicanism in most of England. Wesleyan Methodist missions began during John Wesley's lifetime and had great success over a long period during which Methodism itself divided into a number of sects and established a definite separation from the Church of England.

Poughill Methodist Church

From the early 19th to the mid-20th century Methodism was the leading form of Christianity in Cornwall but it is now in decline. The Church of England was in the majority from the reign of Queen Elizabeth until the Methodist revival of the 19th century: before the Wesleyan missions dissenters were very few in Cornwall. The county remained within the Diocese of Exeter until 1876 when the Anglican Diocese of Truro was created (the first Bishop was appointed in 1877). Roman Catholicism was virtually extinct in Cornwall after the 17th century except for a few families such as the Arundells of Lanherne. From the mid-19th century the church reestablished episcopal sees in England, one of these being at Plymouth. Since then immigration to Cornwall has brought more Roman Catholics into the population.

Physical Geography

Satellite image of Cornwall

Cornwall forms the tip of the south-west peninsula of the island of Great Britain, and is therefore exposed to the full force of the prevailing winds that blow in from the Atlantic Ocean. The coastline is composed mainly of resistant rocks that give rise in many places to impressive cliffs. Cornwall has a border with only one other county, Devon, which is formed almost entirely by the River Tamar and (to the north) by the Marsland Valley.

Coastal Areas

The north and south coasts have different characteristics. The north coast on the Celtic Sea, part of the Atlantic Ocean, is more exposed and therefore has a wilder nature. The prosaically named High Cliff, between Boscastle and St Gennys, is the highest sheer-drop cliff in Cornwall at 223 metres (732 ft).However, there are also many extensive stretches of fine golden sand which form the beaches that are so important to the tourist industry, such as those at Bude, Polzeath, Watergate Bay, Perranporth, Porthtowan, Fistral Beach, Newquay, St Agnes, St Ives, and on the south coast Gyllyngvase beach in Falmouth and the large beach at Praa Sands further to the south west. There are two river estuaries on the north coast: Hayle Estuary and the estuary of the River Camel, which provides Padstow and Rock with a safe harbour. The seaside town of Newlyn is a popular holiday destination, as it is one of the last remaining traditional Cornish fishing ports, with views reaching over Mount›s Bay.

St Michael's Mount in Marazion

The south coast, dubbed the "Cornish Riviera", is more sheltered and there are several broad estuaries offering safe anchorages, such as at Falmouth and Fowey. Beaches on the south coast usually consist of coarser sand and shingle, interspersed with rocky sections of wave-cut platform. Also on the south coast, the picturesque fishing village of Polperro, at the mouth of the Pol River, and the fishing port of Looe on the River Looe are both popular with tourists.

Inland Areas

The interior of the county consists of a roughly east-west spine of infertile and exposed upland, with a series of granite intrusions, such as Bodmin Moor, which contains the highest land within Cornwall. From east to west, and with approximately descending altitude, these are Bodmin Moor, Hensbarrow north of St Austell, Carnmenellis to the south of Camborne, and the Penwith or Land›s End peninsula. These intrusions are the central part of the granite outcrops that form the exposed parts of the Cornubian batholith of south-west Britain, which also includes Dartmoor to the east in Devon and the Isles of Scilly to the west, the latter now being partially submerged.

Cornwall is known for its beaches (Porthcurno beach illustrated) and rugged coastline

The intrusion of the granite into the surrounding sedimentary rocks gave rise to extensive metamorphism and mineralisation, and this led to Cornwall being one of the most important mining areas in Europe until the early 20th century. It is thought tin was mined here as early as the Bronze Age, and copper, lead, zinc and silver have all been mined in Cornwall. Alteration of the granite also gave rise to extensive deposits of China Clay, especially in the area to the north of St Austell, and the extraction of this remains an important industry.

The uplands are surrounded by more fertile, mainly pastoral farmland. Near the south coast, deep wooded valleys provide sheltered conditions for flora that like shade and a moist, mild climate. These areas lie mainly on Devonian sandstone and slate. The north east of Cornwall lies on Carboniferous rocks known as the Culm Measures. In places these have been subjected to severe folding, as can be seen on the north coast near Crackington Haven and in several other locations.

The Lizard Peninsula

The geology of the Lizard peninsula is unusual, in that it is mainland Britain's only example of an ophiolite, a section of oceanic crust now found on land. Much of the peninsula consists of the dark green and red Precambrian serpentinite, which forms spectacular cliffs, notably at Kynance Cove, and carved and polished serpentine ornaments are sold in local gift shops. This ultramafic rock also forms a very infertile soil which covers the flat and marshy heaths of the interior of the peninsula. This is home to rare plants, such as the Cornish Heath, which has been adopted as the county flower.

HILLS AND HIGH POINTS

Ecology

Cornwall has varied habitats including terrestrial and marine ecosystems. One noted species in decline locally is the Reindeer lichen, which species has been made a priority for protection under the national UK Biodiversity Action Plan.

Red-billed chough: *Pyrrhocorax pyrrhocorax pyrrhocorax*. These birds, once commonly seen throughout Cornwall, experienced a severe decline in their population in the 20th century.

Botanists divide Cornwall and Scilly into two vice-counties: West (1) and East (2). The standard flora is by F. H. Davey Flora of Cornwall (1909). Davey was assisted by A. O. Hume and he thanks Hume, his companion on excursions in Cornwall and Devon, and for help in the compilation of that Flora, publication of which was financed by him.

Climate

Cornwall has a temperate Oceanic climate (Köppen climate classification: Cfb) and has the mildest and sunniest climate in the United Kingdom, as a result of its southerly latitude and the influence of the Gulf Stream. The average annual temperature in Cornwall ranges from 11.6 °C (52.9 °F) on the Isles of Scilly to 9.8 °C (49.6 °F) in the central uplands. Winters are amongst the warmest in the country due to the southerly latitude and moderating effects of the warm ocean currents, and frost and snow are very rare at the coast and are also rare in the central upland areas. Summers are however not as warm as in other parts of southern England. The surrounding sea and its southwesterly position mean that Cornwall's weather can be relatively changeable.

Cornwall is one of the sunniest areas in the UK, with over 1541 hours of sunshine per year, with the highest average of 7.6 hours of sunshine per day in July. The moist, mild air coming from the south west brings higher amounts of rainfall than in eastern Great Britain, at 1,051 to 1,290 mm (41.4 to 50.8 in) per year, however not as much as in more northern areas of the west coast. The Isles of Scilly, for example, where there are on average less than two days of air frost per year, is the only area in the UK to be in the USDAHardiness zone 10. In Scilly there is on average less than one day of air temperature exceeding 30 °C per year and it is in the AHS Heat Zone 1. Extreme temperatures in Cornwall are particularly rare; however, extreme weather in the form of storms and floods is common.

Politics and Administration

Local politics

Cornwall Council's headquarters in **Truro**

With the exception of the Isles of Scilly, Cornwall is governed by a unitary authority, Cornwall Council, based in Truro. The Crown Court is based at the Courts of Justice in Truro. Magistrates' Courts are found in Truro (but at a different location to the Crown Court), Bodmin, Penzance and Liskeard.

The Isles of Scilly form part of the ceremonial county of Cornwall and have, at times, been served by the same county administration. Since 1890 they have been administered by their own unitary authority, the Council of the Isles of Scilly. They are grouped with Cornwall for other administrative purposes, such as the National Health Service and Devon and Cornwall Police.

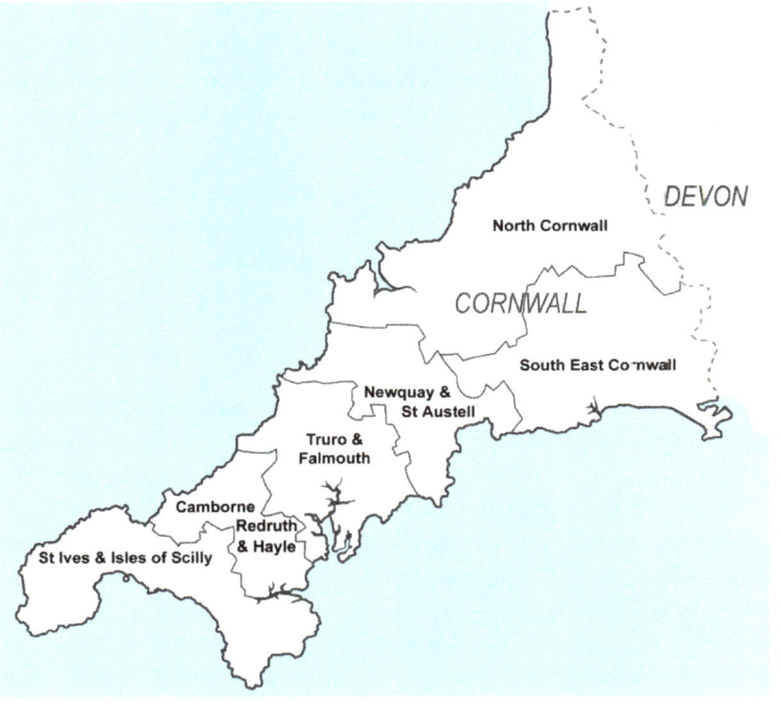

From the 2010 general election, Cornwall has had six parliamentary constituencies

Before reorganisation on 1 April 2009, council functions throughout the rest of Cornwall were organised on a two-tier basis, with a county council and district councils for its six districts, Caradon, Carrick, Kerrier, North Cornwall, Penwith, and Restormel. While projected to streamline services, cut red tape and save around £17 million a year, the reorganisation was met with wide opposition, with a poll in 2008 giving a result of 89% disapproval from Cornish residents.

The first elections for the unitary authority were held on 4 June 2009. The council has 123 seats; the largest party (in 2015) is the Liberal Democrats with 37 seats (37), followed by the Independents, 36 seats (23), Conservative Party with 31 seats (46) Labour Party 8 seats (1), UKIP 6 seats (0), Mebyon Kernow with 4 seats (6), Greens 1 seat, (0). (Number of seats prior to 2013 election in brackets.)

Before the creation of the unitary council, the former county council had 82 seats, the majority of which were held by the Liberal Democrats, elected at the 2005 county council elections. The six former districts had a total of 249 council seats, and the groups with greatest numbers of councillors were Liberal Democrats, Conservatives, and Independents.

Parliament and National Politics

Following a review by the Boundary Commission for England taking effect at the 2010 general election, Cornwall is divided into six county constituencies to elect MPs to the House of Commons of the United Kingdom.

Before the 2010 boundary changes Cornwall had five constituencies all of which were won by Liberal Democrats in the 2005 general election. At the 2010 general election Liberal Democrat candidates won three constituencies and Conservative candidates won three constituencies. At the 2015 general election all six Cornish seats were won by Conservative candidates.

Until 1832, Cornwall had 44 MPs – more than any other county – reflecting the importance of tin to the Crown. Most of the increase in numbers of MPs came between 1529 and 1584 after which there was no change until 1832.

Self-Rule Movement

Cornish nationalists have organised into two political parties: Mebyon Kernow, formed in 1951, and the Cornish Nationalist Party. In addition to the political parties, there are various interest groups such as the Revived Cornish Stannary Parliament and the Celtic League. The Cornish Constitutional Convention was formed in 2000 as a cross-party organisation including representatives from the private, public and voluntary sectors to campaign for the creation of a Cornish Assembly, along the lines of the National Assembly for Wales, Northern Ireland Assembly and the Scottish Parliament. Between 5 March 2000 and December 2001, the campaign collected the signatures of 41,650 Cornish residents endorsing the call for a devolved assembly, along with 8,896 signatories from outside Cornwall. The resulting petition was presented to the Prime Minister, Tony Blair. The Liberal Democrats recognise Cornwall's claims for greater autonomy, as do the Liberal Party.

"The new single council is also the opportunity to gain more control over local issues from regional and national Government bureaucrats – the first step on our way to a Cornish Assembly." – The Liberal Democrat Manifesto for 2009

An additional political issue is the recognition of the Cornish people as a minority.

Cornish National Identity

Cornwall is recognised by several organisations, including the Cornish nationalist party Mebyon Kernow, the Celtic League and the International Celtic Congress, as one of the six Celtic nations, alongside Brittany, Ireland, the Isle of Man, Scotland and Wales. Alongside Asturias and Galicia, Cornwall is also recognised as one of the eight Celtic nations by the Isle of Man Government and the Welsh Government. Cornwall is represented, as one of the Celtic nations, at the Festival Interceltique de Lorient, an annual celebration of Celtic culture held in Brittany.

Cornwall Council consider Cornwall's unique cultural heritage and distinctiveness to be one of the area's major assets. They see Cornwall's language, landscape, Celtic identity, political history, patterns of settlement, maritime tradition, industrial heritage, and non-conformist tradition, to be among the features making up its "distinctive" culture. However, it is uncertain how many of the people living in Cornwall consider themselves to be Cornish; results from different surveys (including the national census) have varied. In the 2001 census, 7 percent of people in Cornwall identified themselves as Cornish, rather than British or English. However, activists have argued that this underestimated the true number as there was no explicit "Cornish" option included in the official census form. Subsequent surveys have suggested that as many as 44 percent identify as Cornish. Many people in Cornwall say that this issue would be resolved if a Cornish option became available on the census. The question and content recommendations for the 2011 Census provided an explanation of the process of selecting an ethnic identity which is relevant to the understanding of the often quoted figure of 37,000 who claim Cornish identity.

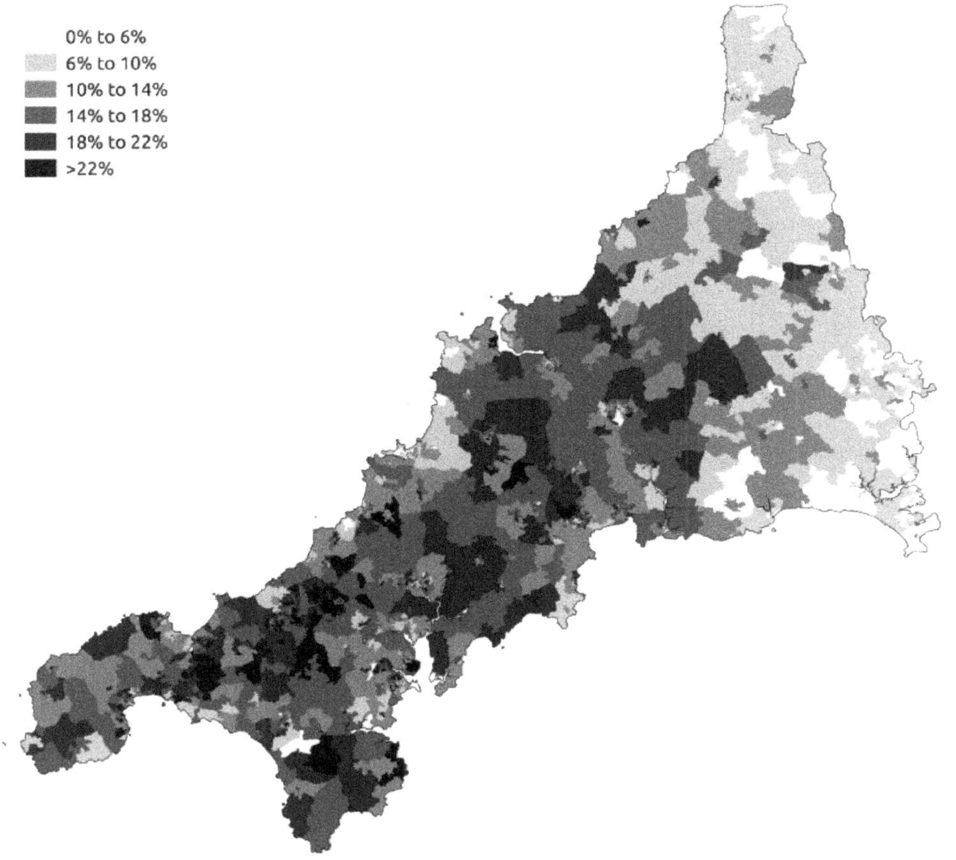

The percentage of respondents who gave "Cornish" as an answer to the National Identity question in the 2011 census.

On 24 April 2014 it was announced that Cornish people would be granted minority status under the European Framework Convention for the Protection of National Minorities.

Settlements and Communication

Truro, Cornwall's administrative centre and only city.

Cornwall's only city, and the home of the council headquarters, is Truro. Nearby Falmouth is notable as a port. St Just in Penwith is the westernmost town in England, though the same claim has been made for Penzance, which is larger. St Ives and Padstow are today small vessel ports with a major tourism and leisure sector in their economies. Newquay on the north coast is famous for its beaches and is a popular surfing destination, as is Bude further north. St Austell is the county›s largest town and more populous than the capital Truro; it was the centre of the china clay industry in Cornwall. Redruth and Camborne form the largest urban area in Cornwall, and both towns were significant as centres of the global tin mining industry in the 19th century (nearby copper mines were also very productive during that period).

Cornwall borders the county of Devon at the River Tamar. Major road links between Cornwall and the rest of Great Britain are the A38 which crosses the Tamar at Plymouth via the Tamar Bridge and the town of Saltash, the A39 road (Atlantic Highway) from Barnstaple, passing through North Cornwall to end in Falmouth, and the A30 which crosses the border south of Launceston crosses Bodmin Moor and connects Bodmin and Truro.

Torpoint Ferry links Plymouth with Torpoint on the opposite side of the Hamoaze. A rail bridge, the Royal Albert Bridge, built by Isambard Kingdom Brunel (1859) provides the only other major transport link. The major city of Plymouth, a large urban centre closest to east Cornwall is an important location for services such as hospitals, department stores, road and rail transport, and cultural venues.

Newquay Cornwall International Airport provides an airlink to the rest of the UK, Ireland and Europe.

Cardiff and Swansea, across the Bristol Channel, have at some times in the past been connected to Cornwall by ferry, but these do not operate currently.

The Isles of Scilly are served by ferry (from Penzance) and by aeroplane, having its own airport — St Mary›s Airport. There are regular flights between St Mary's and Land's End Airport, near St Just, and Newquay Airport; during the summer season, a service also exist between St Mary's and Exeter International Airport, in Devon.

Flag

Souvenir flags outside a Cornish café

Saint Piran's Flag is regarded by many as the national flag of Cornwall, and an emblem of the Cornish people; and by others as the county flag. The banner of Saint Piran is a white cross on a black background (in terms of heraldry ‹sable, a cross argent›). Saint Piran is supposed to have adopted these two colours from seeing the white tin in the black coals and ashes during his supposed discovery of tin. Davies Gilbert in 1826 described it as anciently the flag of St Piran and the banner of Cornwall, and another history of 1880 said that: «The white cross of St. Piran was the ancient banner of the

Cornish people.» The Cornish flag is an exact reverse of the former Breton national flag (black cross) and is known by the same name «Kroaz Du».

There are also claims that the patron saint of Cornwall is Saint Michael or Saint Petroc, but Saint Piran is by far the most popular of the three and his emblem is internationally recognised as the flag of Cornwall. St Piran's Day (5 March) is celebrated by the Cornish diaspora around the world.

Heraldry

For the heraldry of Cornwall see:

- **Cornish heraldry**
- **Cornish corporate heraldry**

Economy

Falmouth Docks is the major port of Cornwall, and one of the largest **natural harbours** in the world

Cornwall is one of the poorest parts of the United Kingdom in terms of per capita GDP and average household incomes. At the same time, parts of the county, especially on the coast, have high house prices, driven up by demand from relatively wealthy retired people and second-home owners. The GVA per head was 65% of the UK average for 2004. The GDP per head for Cornwall and the Isles of Scilly was 79.2% of the EU-27 average for 2004, the UK per head average was 123.0%. In 2011, the latest available figures, Cornwall›s (including the Isles of Scilly) measure of wealth was 64% of the European average per capita.

Historically mining of tin (and later also of copper) was important in the Cornish economy. The first reference to this appears to be by Pytheas: see above. Julius Caesar was the last classical writer to mention the tin trade, which appears to have declined during the Roman occupation. The tin trade revived in the Middle Ages and its importance to the Kings of England resulted in certain privileges being granted to the tinners; the Cornish Rebellion of 1497 is attributed to grievances of the tin miners. In the mid-19th century, however, the tin trade again fell into decline. Other primary industries that have declined since the 1960s include china clay production, fishing and farming.

The **Eden Project** near St Austell, Cornwall's largest tourist attraction in terms of visitor numbers

Today, the Cornish economy depends heavily on its tourist industry, which makes up around a quarter of the economy. The official measures of deprivation and poverty at district and ‹sub-ward' level show that there is great variation in poverty and prosperity in Cornwall with some areas among the poorest in England and others among the top half in prosperity. For example, the ranking of 32,482 sub-wards in England in the index of multiple deprivation (2006) ranged from 819th (part of Penzance East) to 30,899th (part of Saltash Burraton in Caradon), where the lower number represents the greater deprivation.

Cornwall is one of two UK areas designated as ‹less developed regions' which qualify for Cohesion Policy grants from the European Union. It was granted Objective 1 status by the European Commission for 2000 to 2006, followed by further rounds of funding known as ‹Convergence Funding› from 2007 to 2013 and ‹Growth Programme› for 2014 to 2020.

Tourism

Par railway station with a **British Rail Class 43 (HST)** introduced by **British Rail** to the **Cornish Main Line** by the **InterCity** sector.

Tourism is estimated to contribute up to 24% of Cornwall's gross domestic product. In 2011 Tourism brought £1.85 billion into the Cornish economy. Cornwall's unique culture, spectacular landscape and mild climate make it a popular tourist destination, despite being somewhat distant from the United Kingdom's main centres of population. Surrounded on three sides by the English Channel and Celtic Sea, Cornwall has many miles of beaches and cliffs; the South West Coast Path follows a complete circuit of both coasts. Other tourist attractions include moorland, country gardens, museums, historic and prehistoric sites, and wooded valleys. Five million tourists visit Cornwall each year, mostly drawn from within the UK. Visitors to Cornwall are served by airports at Newquay and Exeter, whilst private jets, charters and helicopters are also served by Perranporth airfield; nightsleeper and daily rail services run between Cornwall, London and other regions of the UK. Cornwall has a tourism-based seasonal economy.

Newquay and Porthtowan are popular destinations for surfers. In recent years, the Eden Project near St Austell has been a major financial success, drawing one in eight of Cornwall›s visitors.

Internet

Cornwall is the landing point for one of the world's fastest high-speed transatlantic fibre optic cables, making Cornwall an important hub within Europe's Internet infrastructure. The Superfast Cornwall project completed in 2015, and saw 95% of Cornish houses and businesses connected to a fibre-based broadband network, with over 90% of properties able to connect with speeds above 24Mbit/s.

Other Industries

Redruth Mine in 1890

Other industries are fishing, although this has been significantly re-structured by EU fishing policies (the Southwest Handline Fishermen's Association has started to revive the fishing industry), and agriculture, which has also declined significantly. Mining of tin and copper was also an industry, but today the derelict mine workings survive only as a World Heritage Site. However, the Camborne School of Mines, which was relocated to Penryn in 2004, is still a world centre of excellence in the field of mining and applied geology and the grant of World Heritage status has attracted funding for conservation and heritage tourism. China clay extraction has also been an important industry in the St Austell area, but this sector has been in decline, and this, coupled with increased mechanisation, has led to a decrease in employment in this sector, although the industry still employs around 2,133 people in Cornwall, and generates over £80 Million to the local economy

Demographics

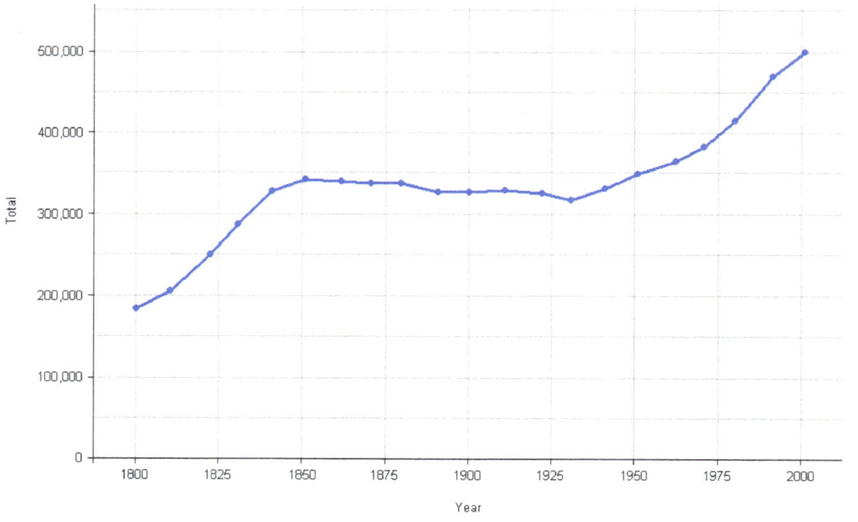

Graph showing Cornwall's population from 1800 to 2000

Cornwall's population was 537,400 at the last census, with a population density of 144 people per square kilometre, ranking it 40th and 41st respectively among the 47 counties of England. Cornwall's population was 95.7% White British and has a relatively high level of population growth. At 11.2% in the 1980s and 5.3% in the 1990s, it had the fifth-highest population growth rate of the English counties. The natural change has been a small population decline, and the population increase is due to inward migration into Cornwall. According to the 1991 census, the population was 469,800.

Cornwall has a relatively high retired population, with 22.9% of pensionable age, compared with 20.3% for the United Kingdom as a whole. This may be due to a combination of Cornwall's rural and coastal geography increasing its popularity as a retirement location, and outward migration of younger residents to more economically diverse areas.

Education System

Cornwall has a comprehensive education system, with 31 state and eight independent secondary schools. There are three further education colleges: Truro and Penwith College, Cornwall College and Callywith College which is due to open in September 2017. The Isles of Scilly only has one school while the former Restormel district has the highest school population, and school year sizes are around 200, with none above 270.

Higher education is provided by Falmouth University, the University of Exeter (including Camborne School of Mines), the Combined Universities in Cornwall, and by Truro College, Penwith College (which combined in 2008 to make Truro and Penwith College) and Cornwall College.

Languages and Dialects

English is the main language used in Cornwall, although the revived Cornish language may be seen on road signs and is spoken fluently by a small minority of people.

Cornish Language

A welcome sign to Penzance, in the English and Cornish languages

The Cornish language is closely related to the other Brythonic languages of Welsh and Breton, and less so to the Goidelic languages of Irish, Scots Gaelic and Manx. The language continued to function visibly as a community language in parts of Cornwall until the late 18th century, and it was claimed in 2011 that the last native speaker did not die until 1914.

There has been a revival of the language since Henry Jenner's Handbook of the Cornish Language was published in 1904. A study in 2000 suggested that there were around 300 people who spoke Cornish fluently. Cornish, however, had no legal status in the UK until 2002. Nevertheless, the language is taught in about twelve primary schools, and occasionally used in religious and civic ceremonies. In 2002 Cornish was officially recognised as a UK minority language and in 2005 it received limited Government funding. A Standard Written Form was agreed in 2008.

Several Cornish mining words are used in English language mining terminology, such as costean, gunnies, vug, kibbal, gossan and kieve.

In the 2010–15 Parliament of the United Kingdom, four Cornish MPs, Andrew George, MP for St Ives, Dan Rogerson, MP for North Cornwall, Steve Gilbert, MP for St Austell and Newquay, and Sarah Newton, MP for Truro and Falmouth repeated their Parliamentary oaths in Cornish.

Culture

Visual Arts

The Tate Gallery at **St Ives**

Artwork in the **Barbara Hepworth** Museum in St Ives

Since the 19th century, Cornwall, with its unspoilt maritime scenery and strong light, has sustained a vibrant visual art scene of international renown. Artistic activity within Cornwall was initially centred on the art-colony of Newlyn, most active at the turn of the 20th century. This Newlyn School is associated with the names of Stanhope Forbes, Elizabeth Forbes, Norman Garstin and Lamorna Birch Modernist writers such as D. H. Lawrence and Virginia Woolf lived in Cornwall between the wars, and Ben Nicholson, the painter, having visited in the 1920s came to live in St Ives with his then wife, the sculptor Barbara Hepworth, at the outbreak of the second world war. They were later joined by the Russian emigrant Naum Gabo,and other artists. These included Peter Lanyon, Terry Frost, Patrick Heron, Bryan Wynter and Roger Hilton. St Ives also houses the Leach Pottery, where Bernard Leach, and his followers championed Japanese inspired studio pottery. Much of this modernist work can be seen in Tate St Ives. The Newlyn Society and Penwith Society of Arts continue to be active, and contemporary visual art is documented in a dedicated online journal.

Music and Festivals

Cornwall has a full and vibrant folk music tradition which has survived into the present and is well known for its unusual folk survivals such as Mummers Plays, the Furry Dance in Helston played by the famous Helston Town Band, and Obby Oss in Padstow.

Newlyn is home to a food and music festival which hosts live music, cooking demonstrations, and displays of locally caught fish.

As in other former mining districts of Britain, male voice choirs and Brass Bands, e.g. Brass on the Grass concerts during the summer at Constantine, are still very popular in Cornwall: Cornwall also has around 40 brass bands, including the six-times National Champions of Great Britain, Camborne Youth Band, and the bands of Lanner and St Dennis.

Cornish players are regular participants in inter-Celtic festivals, and Cornwall itself has several lively inter-Celtic festivals such as Perranporth's Lowender Peran folk festival.

On a more modern note, contemporary musician Richard D. James (also known as Aphex Twin) grew up in Cornwall, as did Luke Vibert and Alex Parks, winner of Fame Academy 2003. Roger Taylor, the drummer from the band Queen was also raised in the county, and currently lives not far from Falmouth. The American singer/songwriter Tori Amos now resides predominantly in North Cornwall not far from Bude with her family. The lutenist, lutarist, composer and festival director Ben Salfield lives in Truro.

Literature

Cornwall's rich heritage and dramatic landscape have inspired writers since at least the 19th century.

Fiction

Sir Arthur Quiller-Couch, author of many novels and works of literary criticism, lived in Fowey: his novels are mainly set in Cornwall. Daphne du Maurier lived at Menabilly near Fowey and many of her novels had Cornish settings, including Rebecca, Jamaica Inn, Frenchman's Creek, My Cousin Rachel, and The House on the Strand. She is also noted for writing Vanishing Cornwall. Cornwall provided the inspiration for The Birds, one of her terrifying series of short stories, made famous as a film by Alfred Hitchcock.

Remains of **Tintagel Castle**, reputedly **King Arthur**'s birthplace

Medieval Cornwall is the setting of the trilogy by Monica Furlong, Wise Child, Juniper, and Colman, as well as part of Charles Kingsley's Hereward the Wake.

Conan Doyle's The Adventure of the Devil's Foot featuring Sherlock Holmes is set in Cornwall. Winston Graham›s series Poldark, Kate Tremayne's Adam Loveday series, Susan Cooper›s novels Over Sea, Under Stone and Greenwitch, and Mary Wesley's The Camomile Lawn are all set in Cornwall. Writing under the pseudonym of Alexander Kent, Douglas Reeman sets parts of his Richard Bolitho and Adam Bolitho series in the Cornwall of the late 18th and the early 19th centuries, particularly in Falmouth.

Hammond Innes›s novel, The Killer Mine; Charles de Lint's novel The Little Country; and Chapters 24 and 25 of J. K. Rowling's Harry Potter and the Deathly Hallows take place in Cornwall (the Harry Potter story at Shell Cottage, which is on the beach outside the fictional village of Tinworth in Cornwall).

David Cornwell, who writes espionage novels under the name John le Carré, lives and writes in Cornwall. Nobel Prize-winning novelist William Golding was born in St Columb Minor in 1911, and returned to live near Truro from 1985 until his death in

1993. D. H. Lawrence spent a short time living in Cornwall. Rosamunde Pilcher grew up in Cornwall, and several of her books take place there.

Poetry

'For The Fallen' plaque with **The Rumps** promontory beyond

The late Poet Laureate Sir John Betjeman was famously fond of Cornwall and it featured prominently in his poetry. He is buried in the churchyard at St Enodoc›s Church, Trebetherick. Charles Causley, the poet, was born in Launceston and is perhaps the best known of Cornish poets. Jack Clemo and the scholar A. L. Rowse were also notable Cornishmen known for their poetry; The Rev. R. S. Hawker of Morwenstow wrote some poetry which was very popular in the Victorian period. The Scottish poet W. S. Graham lived in West Cornwall from 1944 until his death in 1986.

The poet Laurence Binyon wrote "For the Fallen" (first published in 1914) while sitting on the cliffs between Pentire Point and The Rumps and a stone plaque was erected in 2001 to commemorate the fact. The plaque bears the inscription «FOR THE FALLEN / Composed on these cliffs, 1914». The plaque also bears below this the fourth stanza (sometimes referred to as «The Ode») of the poem:

They shall grow not old, as we that are left grow old
Age shall not weary them, nor the years condemn
At the going down of the sun and in the morning
We will remember them

Other Literary Works

Cornwall produced a substantial number of passion plays such as the Ordinalia during the Middle Ages. Many are still extant, and provide valuable information about the Cornish language.

Prolific writer Colin Wilson, best known for his debut work The Outsider (1956) and for The Mind Parasites (1967), lived in Gorran Haven, a small village on the southern Cornish coast. The writer D. M. Thomas was born in Redruth but lived and worked in Australia and the United States before returning to his native Cornwall. He has written novels, poetry, and other works, including translations from Russian.

Thomas Hardy›s drama The Queen of Cornwall (1923) is a version of the Tristan story; the second act of Richard Wagner›s opera Tristan und Isolde takes place in Cornwall, as do Gilbert and Sullivan›s operettas The Pirates of Penzance and Ruddigore. A level of Tomb Raider: Legend, a game dealing with Arthurian Legend, takes place in Cornwall at a museum above King Arthur's tomb.

The fairy tale Jack the Giant Killer takes place in Cornwall.

Sports and Games

As its population is comparatively small, and largely rural, Cornwall's contribution to British national sport in the United Kingdom has been limited; the county's greatest successes have come in fencing. In 2014, half of the men's GB team fenced for Truro Fencing Club, and 3 Truro fencers appeared at the 2012 Olympics. Truro, all of the towns and some villages have football clubs belonging to the Cornwall County Football Association, and the Cornwall County Cricket Club plays as one of the minor counties of English cricket. Viewed as an "important identifier of ethnic affiliation", rugby union has become a sport strongly tied to notions of Cornishness. and since the 20th century, rugby union in Cornwall has emerged as one of the most popular spectator and team sports in Cornwall (perhaps the most popular), with professional Cornish rugby footballers being described as a «formidable force», «naturally independent, both in thought and deed, yet paradoxically staunch English patriots whose top players have represented England with pride and passion». In 1985, sports journalist Alan Gibson made a direct connection between love of rugby in Cornwall and the ancient parish games of hurling and wrestling that existed for centuries before rugby officially began. Among Cornwall›s native sports are a distinctive form of Celtic wrestling related to Breton wrestling, and Cornish hurling, a kind of mediaeval football played with a silver ball (distinct from Irish Hurling). Cornish Wrestling is Cornwall's oldest sport and as Cornwall's native tradition it has travelled the world to places like Victoria, Australia and Grass Valley, California following the miners and gold rushes. Cornish hurling now takes place at St. Columb Major, St Ives, and less frequently at Bodmin. Cornwall is also one of the few places in England where shinty is played; Cornwall Shinty Club was set up in 2012 after the sport was extinct for centuries in the county.

Surfing and other Water Sports

The world pilot gig rowing championships take place annually in the Isles of Scilly.

Cornwall's north coast is known as a centre for **surfing**

Due to its long coastline, various maritime sports are popular in Cornwall, notably sailing and surfing. International events in both are held in Cornwall. Cornwall hosted the Inter-Celtic Watersports Festival in 2006. Surfing in particular is very popular, as locations such as Bude and Newquay offer some of the best surf in the UK. Pilot gig rowing has been popular for many years and the World championships takes place annually on the Isles of Scilly. On 2 September 2007, 300 surfers at Polzeath beach set a new world record for the highest number of surfers riding the same wave as part of the Global Surf Challenge and part of a project called Earthwave to raise awareness about global warming.

Cuisine

Cornwall has a strong culinary heritage. Surrounded on three sides by the sea amid fertile fishing grounds, Cornwall naturally has fresh seafood readily available; Newlyn is the largest fishing port in the UK by value of fish landed, and is known for its wide range of restaurants. Television chef Rick Stein has long operated a fish restaurant in Padstow for this reason, and Jamie Oliver chose to open his second restaurant, Fifteen, in Watergate Bay near Newquay. MasterChef host and founder of Smiths of Smithfield, John Torode, in 2007 purchased Seiners in Perranporth. One famous local fish dish is Stargazy pie, a fish-based pie in which the heads of the fish stick through the piecrust, as though "star-gazing". The pie is cooked as part of traditional celebrations for Tom Bawcock's Eve, but is not generally eaten at any other time.

Cornwall is perhaps best known though for its pasties, a savoury dish made with pastry. Today's pasties usually contain a filling of beef steak, onion, potato and swede with salt and white pepper, but historically pasties had a variety of different fillings. "Turmut, 'tates and mate" (i.e. "Turnip, potatoes and meat", turnip being the Cornish and Scottish term for swede, itself an abbreviation of 'Swedish Turnip', the British term for rutabaga) describes a filling once very common. For instance, the licky pasty contained mostly leeks, and the herb pasty contained watercress, parsley, and shallots. Pasties are often locally referred to as oggies. Historically, pasties were also often made with sweet fillings such as jam, apple and blackberry, plums or cherries. The wet climate and relatively poor soil of Cornwall make it unsuitable for growing many arable crops. However, it is ideal for growing the rich grass required for dairying, leading to the production of Cornwall's other famous export, clotted cream. This forms the basis for many local specialities including Cornish fudge and Cornish ice cream. Cornish clotted cream has Protected Geographical Status under EU law, and cannot be made anywhere else. Its principal manufacturer is A. E. Rodda & Son of Scorrier.

Local cakes and desserts include Saffron cake, Cornish heavy (hevva) cake, Cornish fairings biscuits, figgy 'obbin, Cream tea and whortleberry pie.

There are also many types of beers brewed in Cornwall – those produced by Sharp's Brewery, Skinner's Brewery and St Austell Brewery are the best-known – including stouts, ales and other beer types. There is some small scale production of wine, mead and cider.

BLACK MOUNTAINS, WALES

The Black Mountains (Welsh: Y Mynyddoedd Duon) are a group of hills spread across parts of Powys and Monmouthshire in southeast Wales, and extending across the England–Wales border into Herefordshire. They are the easternmost of the four ranges of hills that comprise the Brecon Beacons National Park, and are frequently confused with the westernmost, which is known as the Black Mountain. The Black Mountains may be roughly defined as those hills contained within a triangle defined by the towns of Abergavenny in the southeast, Hay-on-Wye in the north and the village of Llangors in the west. Other ‹gateway› towns to the Black Mountains include Talgarth and Crickhowell. The range of hills is well known to walkers and ramblers for the ease of access and views from the many ridge trails, such as that on the Black Hill (Herefordshire) at the eastern edge of the massif.

Pen Cerrig-calch and Pen Allt-mawr in the south-west of the Black Mountains range

Name

In his description of a Blak Montayne, the antiquarian John Leland refers to a massif extending between Carmarthen and Monmouth i.e. what is now considered to be the Brecon Beacons in the wider modern sense of that term, thus including the Black Mountain far to the west and the intervening high ground. There is a suggestion too that the names Hatterrall Hill and Mynydd y Gader may also once have been used to apply to the entire range of the Black Mountains though the former later became confined to the vicinity of its eastern ridge. The latter is now recognised in Pen y Gader-fawr (Pen y Gadair Fawr) and Gader Fawr (Gadair Fawr); names applied to the hill at SO 229288. Cadair, mutated to ‹gadair' and anglicised as ‹gader', means ‹seat' or ‹chair' in Welsh.

Mountains

The summit of **Black Mountain** crossed by the **Offa›s Dyke Path**

The southern part of the **Grwyne Fechan** valley in the Black Mountains

Waun Fach from **Mynydd Llysiau**

The highest mountain in the group is Waun Fach whose heavily eroded peat summit plateau attains a height of 811 metres (2,661 ft). Other significant summits towards the northern edge of the range include the 703 metres (2,306 ft) peak of Black Mountain with its northern outpost of Hay Bluff (Welsh: Penybegwn), 677 metres (2,221 ft), Rhos Dirion, 713 metres (2,339 ft) and Lord Hereford's Knob or Twmpa, 690 metres (2,264 ft). Towards the south of the range are the more independent summits of Crug Mawr at 550 metres (1,804 ft), Pen Allt-mawr at 719 metres (2,359 ft) and the 701 metres (2,300 ft) peak of Pen Cerrig-calch which rises prominently above the Usk Valley.

Outlying summits, all of which are classed as Marilyns, include the Sugar Loaf (Welsh: Pen-y-Fal), Mynydd Troed and Mynydd Llangorse. The lower and separate hills of Allt yr Esgair, Myarth, Bryn Arw and Ysgyryd Fawr (also known as ‹The Skirrid›, Skyrrid or ‹Holy Mountain›) are scattered along the southern fringe of the Black Mountains.

In his work People of the Black Mountains, Raymond Williams described the Black Mountains thus:

See this layered sandstone in the short mountain grass. Place your right hand on it, palm downward. See where the summer sun rises and where it stands at noon. Direct your index finger midway between them. Spread your fingers, not widely. You now hold this place in your hand.

The six rivers rise in the plateau towards your wrist. The first river, now called Mynwy, flows at the outside edge of your thumb. The second river, now called Olchon, flows between your thumb and the first finger, to join the Mynwy at the top of your thumb. The third river, now called Honddu, flows between your first and second fingers and then curves to join the Mynwy. The fourth river, now called Grwyne Fawr, flows between your second and third fingers and then curves the other way, south, to join the fifth river, now called Grwyne Fechan, that has been flowing between your third and your outside finger. The sixth river, now called Rhiangoll, flows at the edge of your outside finger.

This is the hand of the Black Mountains, the shape first learned. Your thumb is Crib y Gath. Your first finger is Curum and Hateral. Your second finger is Ffawyddog, with Tal y Cefn and Bal Mawr at its knuckles. Your third finger is Gadair Fawr. Your outside finger is Allt Mawr, from Llysiau to Cerrig Calch and its nail is Crug Hywel. On the high plateau of the back of your hand are Twyn y Llech and Twmpa, Rhos Dirion, Waun Fach and Y Das. You hold their shapes and their names.

Geology

Bedrock

The Black Mountains are composed almost exclusively of rocks assigned to the Old Red Sandstone and dating from the Devonian period. This thick sedimentary sequence comprises sandstones, mudstones, siltstones and numerous thin limestones. The exception is the summit area of Pen Cerrig-calch where a thin sequence of Carboniferous rocks occur, an outlier of the more extensive outcrop to the south of the Usk valley. The lower slopes of these hills are formed from the mudstone-rich St Maughans beds at the top of which lies a calcrete - a discontinuous limestone band known as the Ffynnon Limestone. Above this are the sandstone-dominated Senni Beds which form the upper reaches of much of the range. Higher again are the Brownstones which form the summit areas of the central and southern parts of the range.

The Old Red Sandstone extends back into the late Silurian period and forward into the earliest part of the Carboniferous period. The body of rock, or facies, is dominated by alluvial sediments and conglomerates at its base, and progresses to a combination of dunes, lakes and river sediments. The familiar red colour of these rocks arises from the presence of iron oxide but not all the Old Red Sandstone is red or sandstone — the sequence also includes conglomerates, mudstones, siltstones and thin limestones and colours can range from grey and green through red to purple.

Glacial Legacy

The area lay at the margins of the British ice-sheet during the ice ages; these hills were shaped by ice from a source in mid-Wales rather than generating any major glaciers of their own. Non-local rock fragments within the **glacial till** show that Wye valley ice penetrated the Rhiangoll valley from the north, moving over the low col at Pengenffordd. No such evidence has been found in the **Vale of Ewyas** though the profile of this valley strongly suggests the presence of a major glacier.

View up the **Vale of Ewyas**

The large landslip at the north end of the **Skirrid**

The valleys of the Grwyne Fawr and Grwyne Fechan were probably ice-free during the last ice age. One result of the over-steepening of valley sides by glacial action is the suite of landslips affecting the range, notably in the Vale of Ewyas. The most impressive of all is that at Darren and Cwmyoy. Another impressive set of landslip forms can be seen at Black Darren and Red Darren (‹Darren› signifies ‹edge› in Welsh) on the eastern side of the Hatterrall ridge west of Longtown. Another, at the northern end of the Skirrid just to the east of Abergavenny, is perhaps most commonly seen, and the section of the mountain shows the landslip prominently when seen from the north, but is also visible from the south such is the scale of the feature.

Exploitation

Scattered around the range are innumerable small quarries, virtually all of which now lie abandoned, once a source of walling and roofing stone for local use. In places the thin Devonian limestones were worked to feed limekilns for the production of lime for agricultural use and in buildings. Old Red Sandstone has also frequently been used in buildings in Herefordshire, Monmouthshire and the former Brecknockshire (now south Powys) of south Wales.

Activities in the Black Mountains

The area is popular for hillwalkers, mountainbikers and horseriders. The Offa's Dyke National Trail runs along the border between England and Wales, whilst both the Beacons Way and the Marches Way also pass through the Black Mountains. The Three Rivers Ride runs along the northern slopes of the massif. The range›s great northern escarpment offers excellent opportunities for gliding, hang gliding and paragliding as winds are forced up and over the hills.

Local Attractions

Llanthony Priory

Bookshop in Hay-on-Wye

St Mary's Chapel at **Capel-y-ffin**

There are several attractive villages and hamlets in this area. The Skirrid Mountain Inn also has a wider reputation, being claimed as the oldest public house in Wales and mentioned in records from AD 1100. Antiquities include Llanthony Priory in the Vale of Ewyas, ruined Craswall Priory, Tretower Castle, Tretower Court, the Iron Age hill fort of Crug Hywel, and the remains of Castell Dinas, an 11th to 13th century castle built on the site of an Iron Age hillfort between Talgarth and Crickhowell. Cwmyoy and Partrishow churches are also worth visiting. The youth hostel at Capel-y-ffin closed in late 2007. The town of books or Hay-on-Wye lies just to the north. It was the first booktown to be established and there are more than two dozen second-hand bookshops to explore. Some specialise in specific subjects, so there is a wide choice for browsers.

Towns and Villages

Settlements in and around the Black Mountains include Hay-on-Wye, Llangors, Talgarth, Crickhowell, Cwmdu, each in Powys, Abergavenny in Monmouthshire and Longtown in Herefordshire. Many act as bases for accessing the hills all year round.

Coordinates: 51°57′N 3°6′W

Cultural Associations

The controversial artist and typeface designer Eric Gill lived at Capel-y-ffin between 1924 and 1928. The artist and poet David Jones worked in the area during the same period. The Cultural Marxist Raymond Williams was born here, and draws on the region to frame his analysis of pastoral poetics and development in The Country and the City.

BRECON BEACONS

The Brecon Beacons (Welsh: Bannau Brycheiniog) is a mountain range in South Wales. In a narrow sense, the name refers to the range of Old Red Sandstone peaks which lie to the south of Brecon. Sometimes referred to as "the central Beacons" they include South Wales' highest mountain, Pen y Fan. The range forms the central section of the Brecon Beacons National Park (Parc Cenedlaethol Bannau Brycheiniog), a designation which also encompasses ranges both to the east and the west of "the central Beacons". This much wider area is also commonly referred to as "the Brecon Beacons", and it includes the Black Mountains to the east as well as the similarly named but quite distinct Black Mountain to the west. The highest peaks include Fan Brycheiniog to the west and Pen y Fan in the central part. They share the same basic geology as the central range, and so exhibit many similar features, such the north-facing escarpment and glacial features such as lakes and cwms below the escarpment. They all fall within the border of the national park.

Description

The Brecon Beacons, seen from the south

Pen y Fan seen from **Cribyn**

Fan Brycheiniog the highest peak on the **Black Mountain**

The Brecon Beacons range, in its narrower sense comprises six main peaks: from west to east these are: Corn Du (873 m or 2864 feet), Pen y Fan, the highest peak (886 m or 2907 feet), Cribyn (795 m or 2608 feet), Fan y Big (719 m or 2359 feet), Bwlch y Ddwyallt (754 m or 2474 ft) and Waun Rydd (769 m or 2523 ft). These summits form a long ridge, and the sections joining the first four form a horseshoe shape around the head of the Taf Fechan river, which flows away to the southeast. To the northeast of the ridge, interspersed with long parallel spurs, are four round-headed valleys or cwms; from west to east these are Cwm Sere, Cwm Cynwyn, Cwn Oergwm and Cwm Cwareli.

The Brecon Beacons are said to be named after the ancient practice of lighting signal fires (beacons) on mountains to warn of attacks by invaders,

The round of the Taf Fechan skyline forms a ridge walk commonly known as the Beacons Horseshoe.

The area was inhabited during the Neolithic times and the succeeding Bronze Age, the most obvious legacy of the latter being the numerous burial cairns which adorn the hills of the centre and west of the National Park. There are especially good examples of round barrows on Fan Foel, Pen y Fan and Corn Du. The former was excavated in 2002-4 and the ashes in the central cist dated to about 2000 BC using carbon dating. A wreath of Meadowsweet was likely placed in the burial.

Over twenty hill-forts were established in the area during the Iron Age. The largest, and indeed the largest in south Wales, were the pair of forts atop Y Garn Goch near Bethlehem - Y Gaer Fawr and Y Gaer Fach - literally ‹the big fort› and ‹the little fort›. The forts are thought to have once been trading and political centres.

When the Romans came to Wales in 43 AD, they stationed more than 600 soldiers in the area. Y Gaer, near the town of Brecon was their main base. During the Norman Conquest many castles were erected throughout the park. One of the most famous is Carreg Cennen Castle but there are many more. Brecon Castle is of Norman origin.

There are many old tracks which were used over the centuries by drovers to take their cattle and geese to market in England. The drovers brought back gorse seed which they sowed to provide food for their sheep.

The area played a significant role during the Industrial Revolution as various raw materials including limestone, silica sand and ironstone were quarried for transport southwards to the furnaces of the industrialising South Wales Valleys.

The Brecon Beacons National Park

The Brecon Beacons are one of four ranges of mountains and hills in South Wales which make up the Brecon Beacons National Park. the National Park was established in 1957, the third of the three Welsh parks after Snowdonia in 1951 and the Pembrokeshire Coast in 1952.

A **Welsh mountain pony**

Mountain Rescue

The Beacons in **winter**

Mountain rescue in south Wales is provided by five volunteer groups, with the police having overall command. In serious situations they are aided by RAF helicopters from RAF Chivenor or RAF Valley. The five groups are:

- (CBMRT)Central Beacons Mountain Rescue Team
- (BMRT) Brecon Mountain Rescue Team
- (LMRT) Longtown Mountain Rescue Team based in the east
- (WBMSART) Western Beacons Mountain Search and Rescue Team
- (SARDA) South Wales – Search and Rescue Dog Association covering South and & Mid Wales

The groups are funded primarily by donations. Their work is not restricted to mountain rescue – they frequently assist the police in their search for missing or vulnerable people in the community.

*View north into **Cwm Llwch** from **Corn Du**, in the Brecon Beacons range*

Military Training

The Brecon Beacons, are used for training members of the UK armed forces and military reservists. The Army's Infantry Battle School is located at Brecon, and the Special Air Service (SAS) use the area to test the fitness of applicants. In July 2013 three soldiers died from overheating or heatstroke on an SAS selection exercise. An army captain had been found dead on Corn Du earlier in the year after training in freezing weather for the SAS.

DARTMOOR

Dartmoor is an area of moorland in southern Devon, England. Protected by National Park status as Dartmoor National Park, it covers 954 km2 (368 sq mi).

The granite which forms the uplands dates from the Carboniferous Period of geological history. The moorland is capped with many exposed granite hilltops known as tors, providing habitats for Dartmoor wildlife. The highest point is High Willhays, 621 m (2,037 ft) above sea level. The entire area is rich in antiquities and archaeology.

Dartmoor is managed by the Dartmoor National Park Authority, whose 22 members are drawn from Devon County Council, local district councils and Government.

Parts of Dartmoor have been used as military firing ranges for over 200 years. The public is granted extensive land access rights on Dartmoor (including restricted access to the firing ranges) and it is a popular tourist destination.

Physical Geography

Geology

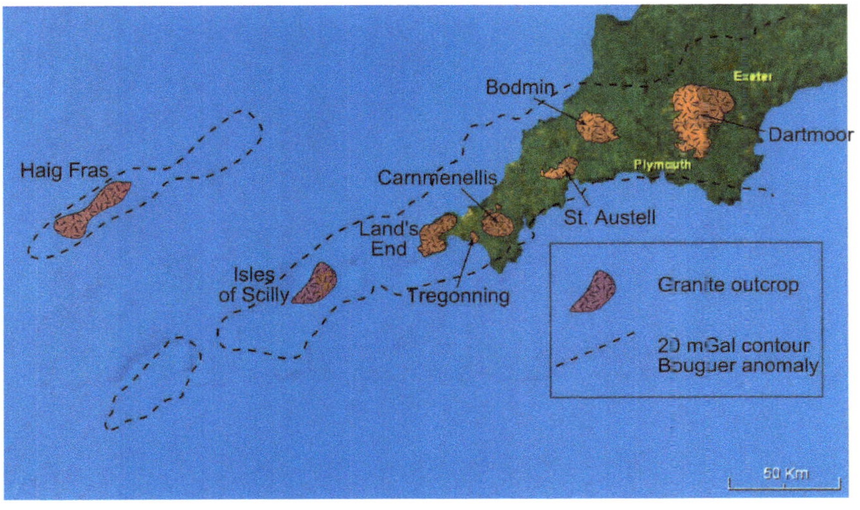

Map showing the main granite outcrops of the Cornubian batholith in southwest England and the gravity anomaly associated with it.

Dartmoor includes the largest area of granite in Britain, with about 625 km^2(241 sq mi) at the surface, though most of it is under superficial peat deposits. The granite (or more specifically adamellite) was intruded at depth as a pluton into the surrounding sedimentary rocks during the Carboniferous period, probably about 309 million years ago. It is generally accepted that the present surface is not far below the original top of the pluton; evidence for this includes partly digested shale xenoliths, contamination of the granite and the existence of two patches of altered sedimentary rock on top of the granite. A considerable gravity anomaly is associated with the Dartmoor pluton as with other such plutons. Measurement of the anomaly has helped to determine the likely shape and extent of the rock mass at depth.

Tors

Panorama of some better known Dartmoor tors in snow

High Willhays with Yes Tor behind

Dartmoor is known for its tors – hills topped with outcrops of bedrock, which in granite country such as this are usually rounded boulder-like formations. More than 160 of the hills of Dartmoor have the word *tor* in their name but quite a number do not. However this does not appear to relate to whether or not there is an outcrop of rock on their summit. The tors are the focus of an annual event known as the Ten Tors Challenge, when around 2400 people aged between 14 and 19 walk for distances of 56, 72 or 88 km (35, 45 or 55 mi) between ten tors on many differing routes.

The highest points on Dartmoor are on the northern moor: High Willhays, 621 m (2,037 ft), (grid referenceSX580895) and Yes Tor, 619 m (2,031 ft), (grid reference SX581901) The highest points on the southern moor are Ryder's Hill, 515 m (1,690 ft), (grid reference SX660690), Snowdon 495 m (1,624 ft), (grid reference SX668684), and an unnamed point of 493 m (1,617 ft) at (grid reference SX603645). The best-known tor on Dartmoor is Haytor (called Hey Tor by William Crossing), 457 m (1,499 ft), (grid referenceSX757771). For a more complete list see List of Dartmoor tors and hills.

Rivers

The high ground of Dartmoor forms the catchment area for many of Devon's rivers. As well as shaping the landscape, these have traditionally provided a source of power for moor industries such as tin mining and quarrying.

The moor takes its name from the River Dart, which starts as the East Dart and West Dart and then becomes a single river at Dartmeet. It leaves the moor at Buckfastleigh, flowing through Totnes below where it opens up into a long ria, reaching the sea at Dartmouth.

For a full list, expand the *Rivers of Dartmoor* navigational box at the bottom of this page.

River Erme at Ivybridge.

Bogs

Much more rain falls on Dartmoor than in the surrounding lowlands. As much of the national park is covered in thick layers of peat (decaying vegetation), the rain is usually absorbed quickly and distributed slowly, so the moor is rarely dry. In areas where water accumulates, dangerous bogs or mires can result. Some of these, topped with bright green moss, are known to locals as "feather beds" or "quakers", because they can shift (or 'quake') beneath a person's feet. Quakers result from sphagnum moss growing over the water that accumulates in the hollows in the granite.

Aune Mire, the source of the River Avon.

The vegetation of the bogs depends on the type and location. Blanket bog, which forms on the highest land where the rainfall exceeds 2,000 millimetres (79 in) a year, consists mainly of cotton-grass (*Eriophorum* species), sedges (*Carex* and *Rhynchospora*), *Bog Asphodel* and *Common Tormentil*, with *Sphagnum* thriving in the wettest patches. The valley bogs have lush growth of rushes, with sphagnum, cross-leaved heath, sundews and several other species.

Some of the bogs on Dartmoor have achieved notoriety. Fox Tor Mires was supposedly the inspiration for Great Grimpen Mire in Conan Doyle's novel *The Hound of the Baskervilles*, although there is a waymarked footpath across it. Sabine Baring-Gould, in his *Book of Dartmoor* (1900) related the story of a man who was making his way through Aune Mire at the head of the River Avon when he came upon a top-hat brim down on the surface of the mire. He kicked it, whereupon a voice called out: "What be you a-doin' to my 'at?" The man replied, "Be there now a chap under'n?" "Ees, I reckon," was the reply, "and a hoss under me likewise."

Climate

Along with the rest of South West England, Dartmoor has a temperate climate which is generally wetter and milder than locations at similar height in the rest of England. At Princetown, near the centre of the moor at a height of 453 metres (1,486 ft), January and February are the coldest months with mean minimum temperatures around 1 °C (34 °F). July and August are the warmest months with mean daily maxima not reaching 18 °C (64 °F). Compared with Teignmouth, which is on the coast about 22 miles (35 km) to the east, the average maximum and minimum temperatures are 3.0 °C (5.4 °F) and 2.6 °C (4.7 °F) lower respectively, and frost is at least five times as frequent. On the highest ground, in the north of the moor, the growing season is less than 175 days – this contrasts with some 300 days along most of the south coast of the county.

Rainfall tends to be associated with Atlantic depressions or with convection. In summer, convection caused by solar surface heating sometimes forms shower clouds and a large proportion of rainfall falls from showers and thunderstorms at this time of year. The wettest months are November and December and on the highest parts of the moor the average annual total rainfall is over 2,000 millimetres (79 in). This compares with less than 800 millimetres (31 in) in the lower land to the east around the Exe Estuary, which is in the rain shadow of the moor. Due to the influence of the Gulf Stream snowfall is not common, though due to its high altitude it is more vulnerable to snowfall than surrounding regions.

Between 1961 and 1990 Met Office data shows that there was an average of 20 days when snow fell on the moor, and over 40 days a year with hail, which is as high as anywhere else in the country. This results when cold polar maritime air that has travelled over a large expanse of warmer ocean is forced to rise over high country.

When average temperatures at Princetown between 1961 and 2000 are compared, the average annual temperature in the decade 1990–2000 was up by 0.2 °C (0.4 °F) and the late winter temperature increased by 0.5 °C (0.9 °F).

The majority of the prehistoric remains on Dartmoor date back to the late Neolithic and early Bronze Age. Indeed, Dartmoor contains the largest concentration of Bronze Age remains in the United Kingdom, which suggests that this was when a larger population moved onto the hills of Dartmoor. The large systems of Bronze Age fields, divided by reaves, cover an area of over 10,000 hectares (39 sq mi) of the lower moors.

The climate at the time was warmer than today, and much of today's moorland was covered with trees. The prehistoric settlers began clearing the forest, and established the first farming communities. Fire was the main method of clearing land, creating pasture and swidden types of fire-fallow farmland. Areas less suited for farming tended to be burned for livestock grazing. Over the centuries these Neolithic practices greatly expanded the upland moors, and contributed to the acidification of the soil and the accumulation of peat and bogs.

After a few thousand years the mild climate deteriorated leaving these areas uninhabited and consequently relatively undisturbed to the present day. The highly acidic soil has ensured that no organic remains have survived, but the durability of the granite has meant that the remains of buildings, enclosures and monuments have survived well, as have flint tools. It should be noted that a number of remains were "restored" by enthusiastic Victorians and that, in some cases, they have placed their own interpretation on how an area may have looked.

Standing Stones

Beardown Man, Dartmoor.

Numerous prehistoric menhirs (more usually referred to locally as standing stones or longstones), stone circles, kistvaens, cairns and stone rows are to be found on the moor. The most significant sites include:

- Upper Erme stone row is the longest on Dartmoor and in fact in the world at 3,300 m (10,800 ft)
- Beardown Man, near Devil's Tor – isolated standing stone 3.5 m (11 ft 6 in) high, said to have another 1 m (3 ft 3 in) below ground. grid reference SX596796
- Challacombe, near the prehistoric settlement of Grimspound — triple stone row. grid reference SX689807
- Drizzlecombe, east of Sheepstor village – stone rows, standing stones, kistvaens and cairns. grid referenceSX591669
- Grey Wethers, near Postbridge — double circle, aligned almost exactly north south. grid reference SX638831
- Laughter Tor, near Two Bridges — standing stone 2.4 m (7 ft 10 in) high and two double stone rows, one 164 m (538 ft) long. grid reference SX652753
- Merrivale, between Princetown and Tavistock — includes a double stone row 182 m (597 ft) long, 1.1 m (3 ft 7 in) wide, aligned almost exactly east-west), stone circles and a kistvaen. grid reference SX554747
- Scorhill (pronounced 'Scorill'), west of Chagford — circle, 26.8 m (88 ft) in circumference, and stone rows. grid reference SX654873
- Shovel Down, north of Fernworthy reservoir — double stone row approximately 120 m (390 ft) long. grid reference SX660859
- Yellowmead Down, a quadruple concentric stone circle and stone rows.

Hut Circles and Kistvaens

There are also an estimated 5,000 hut circles still surviving although many have been raided over the centuries by the builders of the traditional dry stone walls. These are the remnants of Bronze Age houses. The smallest are around 1.8 m (6 ft) in diameter, and the largest may be up to five times this size.

Some have L-shaped porches to protect against wind and rain; some particularly good examples are to be found at Grimspound. It is believed that they would have had a conical roof, supported by timbers and covered in turf or thatch.

There are also numerous kistvaens, Neolithic stone box-like tombs.

The Historical Period

The climate became wetter and cooler over the course of a thousand years from around 1000 BC, resulting in much of high Dartmoor being largely abandoned by its early inhabitants.

It was not until the early Mediaeval period that the weather again became warmer, and settlers moved back onto the moors. Like their ancient forebears, they also used the natural granite to build their homes, preferring a style known as the longhouse — some of which are still inhabited today, although they have been clearly adapted over the centuries. Many are now being used as farm buildings, while others were abandoned and fell into ruin.

The earliest surviving farms, still in operation today, are known as the Ancient Tenements. Most of these date back to the 14th century and sometimes earlier.

Ancient cross close to Crazywell Pool

Some way into the moor stands the town of Princetown, the site of Dartmoor Prison, which was originally built by Isbell Rowe & Company, Plymouth, for prisoners of war from the Napoleonic Wars and the War of 1812. The prison has an incorrect reputation for being escape-proof, due to both the buildings themselves and its physical location.

The Dartmoor landscape is scattered with the marks left by the many generations who have lived and worked there over the centuries – such as the remains of the Dartmoor tin-mining industry, and farmhouses long since abandoned. Indeed, the industrial archaeology of Dartmoor is a subject in its own right.

Myths and Literature

Dartmoor is known for its myths and legends. It is reputedly the haunt of pixies, a headless horseman, a mysterious pack of "spectral hounds", and a large black dog, among others. During the Great Thunderstorm of 1638, the moorland village of Widecombe-in-the-Moor was even said to have been visited by the Devil.

Many landmarks have ancient legends and ghost stories associated with them, such as the allegedly haunted Jay's Grave, the ancient burial site of Childe's Tomb, the rock pile called Bowerman's Nose, and the stone crosses that mark former mediaeval routes across the moor.

A few stories have emerged in recent decades, such as the "hairy hands", that are said to attack motorists on the B3212 near Two Bridges; and the "Beast of Dartmoor", a supposed big cat.

Dartmoor has inspired a number of artists and writers, such as Sir Arthur Conan Doyle in *The Hound of the Baskervilles* and *The Adventure of Silver Blaze*, R. D. Blackmore, Eden Phillpotts, Beatrice Chase, Agatha Christie, Rosamunde Pilcher, and the Reverend Sabine Baring-Gould. In *Harry Potter and the Goblet of Fire*, the fictional 1994 Quidditch World Cup final between Ireland and Bulgaria was hosted on the moor.

Ownership and Access

Over half of Dartmoor National Park (57.3%) is private land; the Forest of Dartmoor being the major part of this, owned by the Duke of Cornwall. The Ministry of Defence owns 14% (see below), 3.8% is owned by water companies (see Dartmoor reservoirs), 3.7% by the National Trust, 1.8% by the Forestry Commission and 1.4% by Dartmoor's national park authority. About 37% of Dartmoor is common land.

Dartmoor differs from some other National Parks in England and Wales, in that since the Dartmoor Commons Act 1985 much of it has been designated as 'Access Land', which, although it remains privately owned, has no restrictions on where walkers can roam. In addition to this Access Land, there are about 730 km (450 mi) of public rights of way on Dartmoor, and many kilometres of permitted footpaths and bridleways where the owner allows access.

Because of the 1985 Act, Dartmoor was largely unaffected by the Countryside and Rights of Way Act 2000, which established similar rights in other rural parts of the country, but in 2006, this Act opened up much of the remaining restricted land to walkers.

Use by the Ministry of Defence

There is a tradition of military usage of Dartmoor dating back to the Napoleonic Wars. Today, a large British Army training camp remains at Okehampton — also the site of an airbase during the Second World War.

The Ministry of Defence (MoD) uses three areas of the northern moor for manoeuvres and live-firing exercises, totalling 108.71 km^2(41.97 sq mi), or just over 11% of the National Park. Red and white posts mark the boundaries of these military areas (shown on Ordnance Survey 1:25,000 scale maps). Flagpoles on many tors in and around the ranges fly red flags when firing is taking place. At other times, members of the public are allowed access. Blank rounds may also be used, but the MoD does not notify the public of this in advance. Some "challenge" and charitable events take place with assistance of the military on Dartmoor including the long established Ten Tors event and the more recent Dartmoor Beast.

Dartmoor's fictional use as an MoD centre for animal testing called Baskerville was referenced in the BBC drama *Sherlock* episode "The Hounds of Baskerville".

The disused Rippon Tor Rifle Range was built to train soldiers during the Second World War, and remained in use until its closure in 1977.

Preservation

The clapper bridge at Postbridge.

Dartmoor Hill pony on Dartmoor.

Throughout human history, the landscape has been exploited for industrial purposes. In recent years, controversy has surrounded the work of industrial conglomerates Imerys and Sibelco (formerly Watts Blake Bearne), who have used parts of the moor for china clay mining. Licences were granted by the British Government but were recently renounced after sustained public pressure from bodies such as the Dartmoor Preservation Association.

The British government has made promises to protect the integrity of the moor; however, the cost of compensating companies for these licences, which may not have been granted in today's political climate, could prove prohibitive.

The military use of the moor has been another source of controversy, such as when training was extended in January 2003. The National Park Authority received 1,700 objections before making the decision. Objectors claimed that Dartmoor should be an area for recreation, and that the training disturbs the peace.

Those who objected included the Open Spaces Society and the Dartmoor Preservation Association. During her lifetime, Lady Sayer was another outspoken critic of the damage which she perceived that the army was doing to the moor.

Towns and Villages

A typical Dartmoor tor close to Haytor.

Dartmoor has a resident population of about 33,000, which swells considerably during holiday periods with incoming tourists. The largest settlements within the National Park are Ashburton (the largest with population about 3,500), Buckfastleigh, Moretonhampstead, Princetown, Yelverton, Horrabridge, South Brent, Christow, and Chagford.

For a full list, expand the *Settlements of Dartmoor* navigational box at the bottom of this page.

Landmarks

- Avon Dam Reservoir – reservoir popular with walks and trout fishing
- Black Tor Beare (Black-a-Tor Copse) – remote woodland of stunted oaks over 305 m (1,001 ft) above sea level in West Okement valley
- Bowerman's Nose – a tor said to be shaped like a nose
- Brentor Church – remote church 344 m (1,129 ft) high on crag at western edge of moor
- Buckfast Abbey – abbey near Buckfastleigh
- Burrator Reservoir – late Victorian reservoir
- Canonteign Falls – second highest waterfall in England
- Castle Drogo – Edwardian faux castle built by Edwin Lutyens on a crag above Teign Gorge, Drewsteignton
- Childe's Tomb – ancient burial site
- Church House, South Tawton – 15th century church ales house
- Cosdon Hill – prominent hill, northern extremity of moor, site of beacon fire for invasion warning
- Cranmere Pool – original letterbox site and location of the legend of Cranmere Binjie
- Crazywell Cross – ancient cross near Crazywell Pool
- Crazywell Pool – artificial lake
- Dartmeet – meeting point of East and West Dart
- Dartmoor Prison – large prison
- Devonport Leat – man-made water channel
- Devonport Leat aqueduct – aqueduct over the River Meavy just down from Raddick Hill Falls
- Duck's Pool – location of a memorial to local writer William Crossing
- Fernworthy Dam & Reservoir – granite-faced concrete dam and lake in Fernworthy Forest, near Chagford
- Fingle Bridge – a 17th-century crossing of the River Teign near Chagford
- Great Links Tor – dominant tor on north west scarp of moor
- Grey Wethers – pair of ancient stone circles
- Grimspound – Bronze Age settlement
- Haytor Granite Tramway – early tramway with stone rails; ran from quarries at Haytor to Stover Canal
- Haytor Rock – prominent tor and viewpoint between Bovey Tracey and Widecombe
- High Willhays – highest point on Dartmoor
- Hound Tor – rugged tor with remains of Iron Age village

- Jay's Grave – burial site
- Lydford Gorge – very deep and narrow gorge with waterfalls
- Meldon Reservoir & viaduct – 201 m (659 ft) wide and 45 m (148 ft) high concrete dam and Victorian iron bridge which itself is 165 m (541 ft) wide and 46 m (151 ft) high
- Nuns Cross – ancient cross
- Powdermills, Cherrybrook – remains of gunpowder grinding mill near Two Bridges
- Rattlebrook railway – remains of horse-worked railway track to peat works near Great Links Tor
- Rippon Tor Rifle Range – disused rifle range
- Scorhill stone circle – (pronounced 'Scorill') well-preserved circle of standing stones near Teign-e-ver, Gidleigh
- Skaigh Valley – narrow, wooded valley leading up on to moor at Belstone; on famous Sticklepath geological fault
- Spitchwick – a swimming spot where the River Webburn joins the River Dart
- Tavistock Canal – 19th-century canal
- Tavy Cleave – river gorge on western moor near Mary Tavy
- Teign-e-ver – confluence of North Teign and Wallabrook; clapper bridges & tolmen stone
- Two Bridges – 18th-century coaching inn
- Warren House Inn – highest inn in south west England
- Wheal Betsy – engine house of mine on Blackdown, Marytavy
- Widgery Cross – granite cross atop Brat Tor, memorial erected by Dartmoor artist William Widgery to commemorate the Golden Jubilee of Queen Victoria in 1887
- Wistman's Wood – remote copse of stunted oaks in valley of West Dart near Two Bridges
- White Hart Hotel – historic 17th-century coaching inn in Moretonhampstead

Leisure Activities

Until the early 19th century Dartmoor was not considered to be a place worth visiting: in the 1540s John Leland wrote in his *Itinerary* that "Dartmore is muche a wilde Morish and forest Ground", and even by 1789 Richard Gough's opinion was that it is a "dreary mountainous tract". At the turn of the 19th century John Swete was one of the first people to visit Dartmoor for pleasure and his journals and watercolour paintings now provide a valuable historical resource.

The oldest leisure pursuit on the moor is hill walking. William Crossing's definitive *Guide to Dartmoor* was published in 1909, and in 1938 a plaque and letterbox in his memory were placed at Duck's Pool on the southern moor. Parts of the Abbots Way, Two Moors Way and the Templer Way are on Dartmoor.

Letterboxing originated on Dartmoor in the 19th century and has become increasingly popular in recent decades. Watertight containers, or ‹letterboxes', are hidden throughout the moor, each containing a visitor's book and a rubber stamp. Visitors take an impression of the letterbox's rubber stamp as proof of finding the box and record their visit by stamping their own personal stamp in the letterbox's logbook. A recent related development is geocaching. Geocache clues make use of GPS coordinates, whereas letterboxing clues tend to consist of grid references and compass bearings.

Whitewater kayaking and canoeing are popular on the rivers due to the high rainfall and their high quality, though for environmental reasons access is restricted to the winter months. The River Dart is the most prominent meeting place, the section known as the Loop being particularly popular. Other white water rivers are the Erme, Tavy, Plym and Meavy.

Other activities are rock climbing on the granite tors and outcrops, some of the well-known venues being Haytor, Hound Tor and The Dewerstone; horse riding, which can be undertaken on any of the common land; cycling (but not on open moorland); and angling for wild brown trout, sea trout and salmon—although much of the river fishing on Dartmoor is privately owned, permits are available for some stretches.

Visitor Centres

The park's main visitor centre is located in Princetown and features exhibits about Dartmoor's history, culture and wildlife, as well as changing displays of local art. The visitor centres located in Postbridge and Haytor feature information, maps, guidebooks and items for exploring the area.

REFERENCES

1. Rackham, Oliver (1986). The History of the Countryside. Dent. ISBN 978-1-84212-440-6.
2. Birks and Madsen (1979). Journal of Ecology, 67.
3. Turner and Hodgson (1979). Journal of Ecology, 67.
4. Camilla Bonn (1998). 'That Jack Cunningham wants half of us out of farming', in Country Life, 15 January 1998, pp. 28–35.
5. McDermot (1911). The History of the Forest of Exmoor.
6. Gimingham (1972). Ecology of heathlands. Chapman & Hall, London.
7. Norton Anthology of English Literature; Romantic Literature.
8. Bleaklow Head at www.hill-bagging.co.uk. Retrieved 11 March 2016
9. Pilton, Barry (1986). One Man and his Bog. Corgi Books. p. 26. ISBN 0-552-12796-5.
10. "Moors for the Future - MoorLIFE Project Sites - Bleaklow". Retrieved 17

August 2015.
11. "NCA Profile: 51 Dark Peak (NE378)". Natural England. Retrieved 4 April 2013.
12. "The Dark Peak (SSSI citation)" (PDF). Natural England. 8 July 1993. Retrieved 4 April 2013.
13. "Eastern Peak District Moors (SSSI citation)" (PDF). Natural England. 22 December 1999. Retrieved 4 April 2013.
14. SAC Standard Data Form
15. "GHOST PLANES OF THE PEAK DISTRICT OF DERBYSHIRE". sheffieldontheinternet.co.uk
16. Place-names in the Standard Written Form (SWF) : List of place-names agreed by the MAGA Signage Panel Archived 15 May 2013 at the Wayback Machine.. Cornish Language Partnership.
17. Pounds, Norman John Greville (2000). A History of the English Parish: the culture of religion from Augustine to Victoria. Cambridge University Press. p. 593. ISBN 978-0-521-63351-2.; p. 72
18. Charoy, B (1986). "Genesis of the Cornubian Batholith (South West England): the example of the Carnmenellis Pluton". Journal of Petrology; Oxford: OUP. Retrieved 27 November 2014.
19. "GENUKI: Cornwall". Genuki.org.uk. Retrieved 27 November 2014.
20. "Bodmin and Launceston". Sheet 186, One-inch map of England and Wales. Ordnance Survey.
21. "Bodmin Moor: Landscape and features". South Penquite Farm. Retrieved 13 August 2016.
22. "The Bodmin Moor Pages ~ The History". Bodminmoor.co.uk. Retrieved 27 November 2014.
23. "Bodmin Moor, North" (PDF). Natural England. 1991. Archived from the original (PDF) on 24 October 2012. Retrieved 26 October 2011.
24. Archived 27 January 2013 at the Wayback Machine.
25. "Bodmin Moor". Important Bird Areas factsheet. BirdLife International. 2013. Archived from the originalon 10 July 2007. Retrieved 2013-04-19.
26. Archived 7 October 2014 at the Wayback Machine.
27. "Cornwall Rivers Project - Geography - Fowey and Lerryn". Cornwallriversporject/org.uk. Retrieved 27 November 2014.
28. "The Rivers of Bodmin Moor". South Penquite Farm. Retrieved 13 August 2016.
29. "Cornwall Rivers Project - Geography - Camel and Allen". Cornwallriversproject.org.uk. Retrieved 27 November 2014.
30. "The Rivers of Bodmin Moor - The Bodmin Moor Pages". Bodminmoore.co.uk. Retrieved 27 November 2014.
31. Bere, Rennie (1982) The Nature of Cornwall. Buckingham: Barracuda Books, pp. 63-67

32. Tilley, C. (1996). World Archaeology: The Power of Rocks: landscape and topography on Bodmin Moor. pp. 151–176.
33. "Programmes - All - Channel 4". Channel 4. Retrieved 27 November 2014.
34. "Bodmin Moor, Cornwall". Channel 4: Time Team. 8 April 2007. Retrieved 9 November 2009.
35. "Secret cornwall - Bodmin moor and its environs". Whitedragon.org.uk. Retrieved 27 November 2014.
36. Clifton-Taylor, A. "Building materials" in: Pevsner, N. (1970) Cornwall. 2nd ed. Penguin Books, p. 29-34
37. "County: Cornwall. Description of Courts: Manorial Courts. Places: Foweymore (Foymore)...". heNational Archives. Retrieved 13 August 2016.
38. Weatherhill, Craig (2009) A Concise Dictionary of Cornish Place-names. Westport, co. Mayo: Evertype; p. 6
39. Bonney, D., Johnson, N., Rose, P. (2008) "Bodmin Moor An archaeological survey Volume 1: The human Landscape c.1800" English Heritage.
40. Giles, C., Herring, P., Johnson, N., Sharpe, A., Smith, J. (2008) "Bodmin Moor An Archaeological survey Volume 2: The industrial and post-medieval landscapes" English Heritage.
41. "The Murder of Charlotte Dymond". Parmaq.com. Retrieved 27 November 2014.
42. Cornish Archaeology; No 34, 1995
43. "The Beast of Bodmin Moor". Natural History Museum. Archived from the original on 7 April 2015. Retrieved 23 April 2014.
44. "Cornwall: definition of Cornwall in Oxford dictionary (American English)". Oxforddictionaries.com. Retrieved 2 November 2015.
45. International Hydrographic Organization
46. "Population estimates for UK mid-2014 analysis tool" (zip). Office for National Statistics. 25 June 2015. Retrieved 27 June 2015.
47. "UK Standard Area Measurements (SAM)" (ZIP). National Statistics Online. Office for National Statistics. July 2007. Retrieved 1 April 2009.
48. Office for National Statistics, Key Figures for 2011 Census: Key Statistics
49. "Data from the 2011 Census (Office for National Statistics)". Cornwall Council. Retrieved 15 November 2013.
50. Stenton, F. M. (1947) Anglo-Saxon England. Oxford: Clarendon Press; p. 337
51. "Blair gets Cornish assembly call". BBC. 11 December 2001. Retrieved 22 May 2008.
52. "Cornish people granted minority status within UK". BBC. 24 April 2014. Retrieved 24 April 2014.
53. Payton, Philip (2004). Cornwall: a history (revised ed.). Fowey: Cornwall Editions Ltd. ISBN 1-904880-00-2.
54. Charles Thomas. (1986). Celtic Britain. (Ancient Peoples & Places Series.)

London: Thames & Hudson

55. Ekwall, E., The Concise Dictionary of English Place-names, 2nd ed., 1940, p. 117a.
56. Watts, Victor (2010). The Cambridge Dictionary of English Place-names (1st paperback ed.). Cambridge University Press. p. 158. ISBN 978-0-521-16855-7.
57. Davies, John (1994). A History of Wales. London: Penguin. p. 69. ISBN 978-0-14-028475-1.
58. Payton (2004), p. 50.
59. "Kingdoms of British Celts – Cornubia". The History Files. Retrieved 1 December 2007.
60. Deacon, Bernard (2007). A Concise History of Cornwall. University of Wales Press. p. 4. ISBN 978-0-7083-2032-7.
61. Cunliffe, Karl, Guerra, McEvoy, Bradley; Oppenheimer, Rrvik, Isaac, Parsons, Koch, Freeman and Wodtko (2010). Celtic from the West: Alternative Perspectives from Archaeology, Genetics, Language and Literature. Oxbow Books and Celtic Studies Publications. p. 384. ISBN 978-1-84217-410-4.
62. Cunliffe, Barry (2009). "A Race Apart: Insularity and Connectivity". Proceedings of the Prehistoric Society. The Prehistoric Society. 75: 55–64.
63. Payton (2004), p. 40.
64. Halliday (1959), p. 51.
65. Halliday (1959), p. 52.
66. Champion, Timothy. "The appropriation of the Phoenicians in British imperial ideology". Nations and Nationalism. 7 (4): 451–465. doi:10.1111/1469-8219.00027.
67. "AD 500 – Tintagel". Archaeology.co.uk. 24 May 2007. Retrieved 24 November 2013.
68. "Medieval Sourcebook: The Annales Cambriae 447–954 (The Annals of Wales)". Fordham.edu. Retrieved 2 November 2015.
69. Weatherhill, Craig Cornovia; p. 10
70. "The Foundation of the Kingdom of England". Third-millennium-library.com. Retrieved 25 September 2010.
71. Annales Cambriae
72. Keynes, Simon; Lapidge, Michael (tr.) (1983), Alfred the Great: Asser's Life of King Alfred and other contemporary sources, London, Penguin Books, p. 175; cf. ibid, p. 89
73. Cornish (and Other) Personal Names from the 10th Century Bodmin Manumissions by Heather Rose Jones
74. "Welsh Journals Online". Welshjournals.llgc.org.uk. Retrieved 2 November 2015.
75. EMR Ditmas, Tristan and Iseult Twelfth Century Romance by Beroul retold from Norman French 1969

76. Williams, Ann & Martin, G. H. (2002) (tr.) Domesday Book: a complete translation, London: Penguin, pp. 341–357

77. Payton (2004), chapter 5.

78. Matthew, David. The Cornish and Welsh Pirates in the Reign of Elizabeth. The English Historical Review Vol. 39, No. 155 (Jul. 1924): Oxford University Press. pp. 337–348.

79. Orme, Nicholas (2000) The Saints of Cornwall

80. "uny". Lelant.info. 5 January 2008. Retrieved 24 November 2013.

81. Doble, G. H. (1960) The Saints of Cornwall. 5 vols. Truro: Dean and Chapter, 1960–70

82. See for example absences from Olsen and Padel's "A tenth century list of Cornish parochial saints" in Cambridge Medieval Celtic Studies; 12 (1986); and from Nova Legenda Angliae by John Capgrave (mid-15th century)

83. "St. Piran – Sen Piran". St-Piran.com. Retrieved 11 May 2007.

84. Henderson, Charles (1935) "Cornwall and her patron saint", In: his Essays in Cornish History. Oxford: Clarendon Press; pp. 197–201

85. Wyatt, Tim (2 May 2014). "Cornish welcome new status". Church Times. Retrieved 8 May 2014.

86. Caroline Brett, "Petroc (fl. 6th cent.)", Oxford Dictionary of National Biography, Oxford University Press, 2004 accessed 16 December 2008

87. Charles-Edwards, T. (1970) "The Seven Bishop Houses of Dyfed," In: Bulletin of the Board of Celtic Studies, vol. 24, (1970–1972), pp. 247–252.

88. Orme, Nicholas (2007) Cornwall and the Cross. Chichester: Phillimore; p. 29

89. Thorn, Caroline, et al., eds. (1979) Cornwall. Chichester: Phillimore

90. Cornish Church Guide (1925) Truro: Blackford

91. Jenner, Henry (1925) "The Holy Wells of Cornwall". In: Cornish Church Guide. Truro: Blackford; pp. 249–257

92. Quiller-Couch, M. & L. (1894) Ancient and Holy Wells of Cornwall. London: Chas. J. Clark

93. Oliver, George (1846) Monasticon Dioecesis Exoniensis: being a collection of records and instruments illustrating the ancient conventual, collegiate, and eleemosynary foundations, in the Counties of Cornwall and Devon, with historical notices, and a supplement, comprising a list of the dedications of churches in the Diocese, an amended edition of the taxation of Pope Nicholas, and an abstract of the Chantry Rolls with supplement and index. Exeter: P. A. Hannaford, 1846, 1854, 1889

94. Orme, Nicholas (2007) Cornwall and the Cross. Chichester: Phillimore; p. 147

95. "Methodism". Cornish-Mining.org.uk. Retrieved 11 May 2007.

96. Shaw, Thomas (1967) A History of Cornish Methodism. Truro: Bradford Barton

97. "Truro Cathedral website – History page". TruroCathedral.org.uk. Retrieved 11

May 2007.

98. Brown, H. Miles (1976) A Century for Cornwall. Truro: Blackford

99. "Diocese of Plymouth". Retrieved 13 April 2009.

100. "The Official Guide to the South West Coast Path". Southwestcoastpath.com. Archived from the original on 11 June 2011. Retrieved 25 September 2010.

101. "The Cornish Heath – County Flower". Archived from the originalon 30 September 2008. Retrieved 5 February 2016.

102. Price, J. H., Hepton, C. E. L. and Honey, S. I. (1979). The Inshore Benthic Biota of the Lizard Peninsula, south west Cornwall: the marine algae – History; Chlorophyta; Phaeophyta. Cornish Studies; no. 7: pp. 7–37

103. Bere, Rennie (1982) The Nature of Cornwall. Buckingham: Barracuda Books

104. "UK climate information". Met Office. 1 May 2014. Retrieved 2 November 2015.

105. "UK climate information". Met Office. 1 May 2014. Retrieved 2 November 2015.

106. "UK climate information". Met Office. 1 May 2014. Retrieved 2 November 2015.

107. "Isles of Scilly; Cornwall through time". visionofbritain.org.uk. Retrieved 19 January 2007.

108. "Isles of Scilly RD; Cornwall through time". visionofbritain.org.uk. Archived from the original on 6 May 2007. Retrieved 19 January 2007.

109. "About your local police". Devon and Cornwall Police. Archived from the original on 7 October 2009. Retrieved 23 September 2009.

110. "One Cornwall – A unified council for Cornwall". Cornwall County Council. Archived from the original on 1 May 2008. Retrieved 22 May 2008.

111. Archived 11 October 2008 at the Wayback Machine.

112. "Cornwall super-council go-ahead". BBC. 25 July 2007. Retrieved 25 July 2007.

113. "Elections and voting in Cornwall | mudhook | Page 2". Mudhook.wordpress.com. Retrieved 2 November 2015.

114. "British Archaeology, no 30, December 1997: Letters". Britarch.ac.uk. Retrieved 25 September 2010.

115. Rowse, A. L. (1941) Tudor Cornwall. London: Cape; pp. 91–94

116. Cornish Constitutional Convention (3 April 2005). "Campaign for a Cornish Assembly – Senedh Kernow". Cornishassembly.org. Retrieved 25 September 2010.

117. "Cornwall builders". Monday, 28 November 2016

118. "Minutes for the Census Sub-Group Meeting held on 23 November 2006". Central & Local Information Partnership. 9 February 2007. Archived from the original on 16 October 2008. Retrieved 23 May 2008.

119. "Lords Hansard Text for 25 Jan 2011 (pt002)". Hansard. Parliament of the United Kingdom. 25 January 2011. Retrieved 8 February 2011. Cornwall sees itself as the fourth Celtic nation of the United Kingdom; Lord Teverson

120. "Mebyon Kernow – The Party for Cornwall – BETA". Mebyon Kernow website. Mebyon Kernow. 2007. Retrieved 8 February 2011.

121. "The Celtic League". Celtic League website. Celtic League. 2011. Retrieved 8 February 2011.

122. "The International Celtic Congress". International Celtic Congress. 2009. Retrieved 16 February 2010.

123. "Welsh Government: Minister in Paris for launch of Celtic festival". Welsh Government website. Welsh Government. 14 March 2002. Archived from the original on 10 March 2012. Retrieved 8 February 2011.

124. "Isle of Man Post Office Website". Isle of Man Post Office website. Isle of Man Government. 2009. Retrieved 8 February 2011.

125. "Site Officiel du Festival Interceltique de Lorient". Festival Interceltique de Lorient website. Festival Interceltique de Lorient. 4 February 2011. Archived from the original on 5 March 2010. Retrieved 8 February 2011.

126. "Cornwall Council – part three". Cornwall Council website. Cornwall Council. 18 March 2010. Retrieved 10 February 2011.

127. Dugan, Emily (6 September 2009). "The Cornish: they revolted in 1497, now they're at it again". London: Independent (The). Retrieved 14 July 2010.

128. "Welsh are more patriotic". BBC. 3 March 2004. Retrieved 14 July 2010.

129. "Information paper: Recommended questions for the 2009 Census Rehearsal and 2011 Census: National Identity" (PDF). Office for National Statistics. December 2008. p. 32. Retrieved 13 May 2009.

130. "2011 Census; 2011 census questionnaire content: question and content recommendations for 2011; ethnic group prioritisation tool"(PDF). Office for National Statistics. pp. 20–22. Retrieved 15 July 2010.

131. "Fast Ferries connecting Wales & the South West – Severn Link, Fast Ferries Connecting Wales and the South West". Severnlink.com. Retrieved 2 November 2015.

132. Rendle, Phil. "Cornwall – The Mysteries of St Piran" (PDF). Proceedings of the XIX International Congress of Vexillology. The Flag Institute. Retrieved 17 January 2010.

133. "Cross of Saint Piran". Flags of the World (FOTW). Retrieved 17 January 2010.

134. Payton (2004), p. 262.

135. "Cornwall (United Kingdom)". Crwflags.com. Retrieved 25 September 2010.

136. "British Flags (United Kingdom) from The World Flag Database". Flags.net. Retrieved 25 September 2010.

137. Dugan, Emily (27 July 2008). "Cornwall: A land of haves, and have nots". The Independent. London. Retrieved 15 May 2013.

138. ONS December 2006

139. Eurostat

140. "Cornwall economy slips back in fight with Europe's poorest". Western Morning News. 7 May 2014. Retrieved 2 November 2015.

141. Halliday (1959), p. 69.
142. Halliday (1959), p. 182.
143. Archived 8 August 2007 at the Wayback Machine.
144. "Cohesion Policy and the United Kingdom" (PDF). European Union. October 2014. Retrieved 28 March 2016. For 2014–2020, the UK has been allocated around €11.8 billion (current prices) in total Cohesion Policy funding: €2.6 billion for less developed regions (Cornwall and the Isles of Scilly, West Wales and the Valleys).
145. "Impact Analysis: ESF Objective 1 Programme Cornwall and the Isles of Scilly, 2000–2006" (PDF). University of Exeter. South West of England Regional Development Agency. March 2008. Archived from the original (PDF) on 7 April 2016. Retrieved 27 March 2016. Cornwall and the Isles of Scilly (IoS) received Objective One status in 1999, primarily as a consequence of their low Gross Domestic Product (GDP) per head (70.3% of the EU average). This resulted from a myriad of underlying socio-economic problems including the large number of people with relatively low levels of qualifications; with lack of basic skills beyond Level two being a particular problem.
146. "Convergence Programme for Cornwall and the Isles of Scilly Operational Programme 2007–13" (PDF). Department for Communities and Local Government. Retrieved 28 March 2016. The Convergence Programme for Cornwall and the Isles of Scilly will succeed the current Objective One Programme, which has been delivered with great effect by the key stakeholders. A total of €458.1 million has been allocated to the new Programme for 2007–13, which will be complemented by €196 million of Convergence ESF funding to support learning and skills related activities
147. "Cornwall Council – Business – Europe". Cornwall Council. 18 December 2015. Retrieved 28 March 2016. The Cornwall and the Isles of Scilly Growth Programme is worth €603,706,863 (excluding technical assistance) and is made up of two main funding streams: European Regional Development Fund (ERDF) -The ERDF Programme is worth €452,780,147 and makes up 75% of the Programme allocation and European Social Fund (ESF) – The ESF Programme is worth €150,926,716 and makes up 25% of the Programme allocation.
148. "Home Page – Cornwall Council". Db.cornwall.gov.uk. Retrieved 2 November 2015.
149. "Tourist earnings top £9bn in Devon and Cornwall". Western Morning News. 21 December 2012. Retrieved 2 November 2015.
150. Visit Cornwall, 2007. "The total number of visitors to the county includes those on business and visiting relatives.http://www.visitcornwall.com/xsdbimgs/Tourism%20in%20Cornwall%202007.pdf Tourism in Cornwall] " (PDF). Archived from the original(PDF) on 18 January 2012. External link in |title= (help)The total number of visitors to the county includes those on business and visiting relatives. Tourism in Cornwall (216 KB)Archived 18 January 2012 at the Wayback Machine.

151. "Cornwall Visitor Guide". www.campsites.co.uk. Retrieved 28 March 2016.
152. Scottish Executive, 2004. A literature review of the evidence base for culture, the arts and sport policy.
153. Pole, Warren (5 July 2009). "Web in trouble? The hidden cables under a Cornish beach feeding the world's internet | Daily Mail Online". Dailymail.co.uk. Retrieved 2 November 2015.
154. Curtis, Sophie (24 July 2014). "Cornwall claims 'half' of UK's direct fibre broadband links". Daily Telegraph. Retrieved 18 March 2016.
155. "Superfast Cornwall Programme". Superfast Cornwall. Retrieved 18 March 2016.
156. "Line-caught wild bass from Cornwall – South West Handline Fishermen's Association". Linecaught.org.uk. Retrieved 25 September 2010.
157. "Cornwall and West Devon Mining Landscape – UNESCO World Heritage Centre". Whc.unesco.org. Retrieved 2 November 2015.
158. "The University of Exeter – Cornwall Campus – Camborne School of Mines". Uec.ac.uk. Retrieved 25 September 2010.
159. "Home". Cornish-mining.org.uk. 14 September 2010. Retrieved 25 September 2010.
160. Imerys Minerals Ltd (2003) Blueprint: Vision for the Future
161. Office for National Statistics, 2001. Population Change in England by County 1981–2000 Archived 30 June 2006 at the Wayback Machine..
162. Office for National Statistics, 2001. Births, Deaths and Natural Change in Cornwall 1974 – 2001 Archived 29 October 2006 at the Wayback Machine..
163. Office for National Statistics, 1996. % of Population of Pension Age (1996) Archived 30 June 2006 at the Wayback Machine..
164. "Legend of Dolly Pentreath outlived her native tongue". This is Cornwall. 4 August 2011. Retrieved 27 June 2012.
165. Archived 18 March 2006 at the Wayback Machine.
166. "Cornish in United Kingdom". European Commission. Retrieved 11 May 2007.
167. "November 2002 – Cornish gains official recognition". BBC News. 6 November 2002. Retrieved 25 September 2010.
168. "June 2005 – Cash boost for Cornish language". BBC News. 14 June 2005. Retrieved 25 September 2010.
169. "An Outline of the Standard Written Form of Cornish" (PDF). Retrieved 25 September 2010.
170. Dictionary of Mining, Mineral, and Related Terms by American Geological Institute and U S Bureau of Mines; pp. 128, 249 & 613
171. "kibbal - Definition of kibbal - Online Dictionary from Datasegment.com". Onlinedictionary.datasegment.com. Retrieved 24 October 2012.

172. "gossan – definition of gossan by the Free Online Dictionary, Thesaurus and Encyclopedia". Thefreedictionary.com. 21 September 2009. Retrieved 24 October 2012.

173. Hunt, Robert (1887). British Mining: A Treatise on the History, Discovery, Practical Development. London: E. & F. N. Spon. p. 778. Retrieved November 8, 2016. The tin stuff is put into a "kieve" about 3½ feet diameter and 2½ feet deep, and with an equal volume of water is continually stirred with a shovel in one direction until the tin stuff is in a state of suspended motion.

174. "MPs swear Oath of Allegiance in Cornish". Maga Kernow. 24 May 2010. Retrieved 26 June 2010.

175. "Elizabeth Adela Forbes". PenleeHouse.org.uk. Retrieved 11 May 2007.

176. "Samuel John Lamorna Birch". HayleGallery.co.uk. Archived from the original on 28 September 2007. Retrieved 11 May 2007.

177. "Virginia Woolf". The New York Times. Retrieved 11 May 2007.

178. "Ben Nicholson". StormFineArts.com. Archived from the original on 30 December 2006. Retrieved 11 May 2007.

179. "Naum Gabo". Artnet.com. Retrieved 11 May 2007.

180. "Bernard Leach and the Leach Pottery". Studio-Pots.com. Archived from the original on 3 April 2007. Retrieved 11 May 2007.

181. "Tate St Ives". Tate.org.uk. Retrieved 11 May 2007.

182. "art and artists in Cornwall including Cornish galleries". art cornwall .org. Retrieved 25 September 2010.

183. "Newlyn fish festival". Newlynfishfestival.org.uk. Retrieved 2 November 2015.

184. "An Daras Cornish Folk Tradition". An-daras.com. Retrieved 2 November 2015.

185. Blackman, Guy (8 May 2005). "The whole Tori – Music – Entertainment". The Age. Melbourne. Retrieved 25 September 2010.

186. "Daphne du Maurier". DuMaurier.org. Retrieved 11 May 2007.

187. "The Birds". MovieDiva.com. Retrieved 11 May 2007.

188. "The Adventure of the Devil's Foot". WorldwideSchool.org. Retrieved 11 May 2007.

189. "Over Sea, Under Stone". Powell's Books. Retrieved 11 May 2007.

190. "The Killer Mine". BoekBesprekingen.nl. Retrieved 11 May 2007.

191. "The Little Country". Amazon.com. Retrieved 11 May 2007.

192. "Shell Cottage". hp-lexicon.org. Retrieved 11 January 2008.

193. "Le Carré betrayed by 'bad lot' spy Kim Philby". Channel 4 News. London: Channel 4. 12 September 2010. Retrieved 26 May 2011.

194. "Biography of William Golding". William-Golding.co.uk. Archivedfrom the original on 17 March 2007. Retrieved 11 May 2007.

195. "St Enodoc Church". RockInfo.co.uk. Archived from the originalon 13 June 2007. Retrieved 11 May 2007.

196. "William Sydney Graham". CPRW.com. Retrieved 11 May 2007.
197. Clegg 2005, p. 10.
198. Pilnick, Brent (1 June 2012). "London 2012: Truro fencers dominate GB Olympic team". BBC Sport. Retrieved 28 July 2016.
199. Harvey, David (2002). Celtic Geographies: Old Culture, New Times. London: Routledge. p. 221. ISBN 978-0-415-22396-6.
200. Gallagher, Brendan (23 October 2008). "Cornish rugby union celebrate 125 years of pride and passion – but are they the lost tribe?". The Daily Telegraph. UK. Retrieved 7 September 2009.
201. "BBC NEWS, Surfers aim to break world record". BBC News. 2 September 2007. Retrieved 25 September 2010.
202. "Objective One media release". Objectiveone.com. Retrieved 25 September 2010.
203. "Cornish recipe site". Alanrichards.org. 25 February 2008. Retrieved 25 September 2010.
204. Martin, Edith (1929). Cornish Recipes, Ancient & Modern. 22nd edition, 1965.
205. "Official list of British protected foods". Europa.eu.int. 23 February 2009. Archived from the original on 2 June 2005. Retrieved 25 September 2010.
206. Mason, Laura; Brown, Catherine (1999) From Bath Chaps to Bara Brith. Totnes: Prospect Books
207. Pettigrew, Jane (2004) Afternoon Tea. Andover: Jarrold
208. Fitzgibbon, Theodora (1972) A Taste of England: the West Country. London: J. M. Dent
209. Morgan, R. 2005. Place-names of Gwent Gwasg Carreg Gwalch, Llanrwst ISBN 0-86381-956-7
210. Toulmin Smith, Lucy (ed.), The Itinerary in Wales of John Leland in or about the years 1536-1539, Vol. 3 Containing Part 6 (The Itinerary in Wales), with a Map, London, 1906
211. Welsh Academy Dictionary
212. British Geological Survey 1:50,000 scale geological maps (England & Wales series) 214 Talgarth & 232 Abergavenny & associated memoirs
213. Davies, Andy; Whittaker, David (2010), Walking on the Brecon Beacons, Cicerone, p. 139, ISBN 978-1-8528-4554-4
214. Farmer, Ben (14 July 2013). "Two soldiers die in heat while training in Brecon Beacons". The Telegraph. Retrieved 14 January 2015.
215. Farmer, Ben (2 July 2014). "No SAS manslaughter charges over Brecon Beacons deaths". The Telegraph. Retrieved 14 January 2015.
216. "General Information Factsheet". Dartmoor National Park Authority. Archived from the original on 8 September 2008. Retrieved 12 July 2009.
217. Durrance & Laming 1982, pp.86, 101
218. Durrance & Laming 1982, p.88

219. British Geological Survey 1997, Gravity Anomaly Map of Britain, Ireland and Adjacent Areas, Smith & Edwards 1:500k

220. Bradt, Hilary; Booth, Janice (2014). South Devon & Dartmoor. Chalfont St Peter: Bradt. p. 222. ISBN 9781841625522.

221. Ian Mercer: Bogs and Mires of Dartmoor, in Hunt & Wills 1977, p.16.

222. Sandles, Tim. "The Bogs & Mires of Dartmoor". Legendary Dartmoor. Retrieved 12 July 2009.

223. Crossing, W, Crossing's Guide to Dartmoor 1912 edition, published by Western Morning News Co. Ltd.

224. Brunsden, Denys; Gerrard, John (1970). "The Physical Environment of Dartmoor". In Crispin Gill. Dartmoor. A New Study. Newton Abbot: David and Charles. pp. 40–42. ISBN 0-7153-5041-2.

225. Sandles, Tim. "Dartmoor's Notorious Fox Tor Mires". Legendary Dartmoor. Retrieved 7 July 2009.

226. Milton 2006, p.2

227. Webb 2006, p.30

228. Webb 2006, p.31

229. Webb 2006, pp.32–33

230. "Prehistoric Dartmoor". Dartmoor National Park Authority. Retrieved 2 September 2009.

231. Role of anthropogenic fire and in creating moors, and moor-burning in sustaining them, described in Pyne 1997, pp. 348–369.

232. Sandles, Tim. "The Hairy Hands". Legendary Dartmoor. Retrieved 7 August 2010.

233. Sandles, Tim. "The Legendary Beast of Dartmoor". Legendary Dartmoor. Archived from the original on 23 August 2010. Retrieved 7 August 2010.

234. "Dartmoor Commons" (PDF). Dartmoor National Park Authority. Archived from the original (PDF) on 27 February 2008. Retrieved 19 March 2008.

235. Dartmoor Commons Act 1985 on the OPSI website

236. "Public Rights of Way". Dartmoor National Park Authority. Archived from the original on 11 September 2009. Retrieved 14 July 2009.

237. http://www.heritagegateway.org.uk/Gateway/Results_Single.aspx?uid=MDV55113&resourceID=104

238. "ViaMichelin Travel. Panorama - Great Britain: The Michelin Green Guide". Viamichelin.com. 3 April 2008. Retrieved 18 April 2010.

239. "Archived copy" (PDF). Archived from the original (PDF) on 9 January 2009. Retrieved 2009-07-12.

240. Milton 2006, pp.19–20

241. Milton 2006, p.36

242. "Walking For All". Dartmoor National Park Authority. Archived from the original on 18 April 2009. Retrieved 12 July 2009.

243. "South West England". UK Rivers Guidebook. Retrieved 18 April 2010.
244. "Canoeing". Dartmoor National Park Authority. Retrieved 12 July 2009.
245. "Climbing". Dartmoor National Park Authority. Archived from the original on 14 July 2009. Retrieved 10 July 2009.
246. "Horse Riding". Dartmoor National Park Authority. Archived from the original on 17 July 2009. Retrieved 12 July 2009.
247. "Cycling on road and off road". Dartmoor National Park Authority. Archived from the original on 30 April 2009. Retrieved 12 July 2009.
248. "Other Activities". Dartmoor National Park Authority. Archived from the original on 17 July 2009. Retrieved 10 July 2009.

CHAPTER 4

Land Improvement through Pastoral Farming

Chapter Contents

Pastoral farming (also known in some regions as livestock farming or grazing) is a form of agriculture aimed at producing livestock, rather than growing crops. Examples include dairy farming, raising beef cattle, and raising sheep for wool. In contrast, arable farming concentrates on crops rather than livestock. Finally, Mixed farming incorporates livestock and crops on a single farm. Some mixed farmers grow crops purely as fodder for their livestock; some crop farmers grow fodder and sell it to pastoral farmers.

INTRODUCTION

Pastoral farming (also known in some regions as livestock farming or grazing) is a form of agriculture aimed at producing livestock, rather than growing crops. Examples include dairy farming, raising beef cattle, and raising sheep for wool. In contrast, arable farming concentrates on crops rather than livestock. Finally, Mixed farming incorporates livestock and crops on a single farm. Some mixed farmers grow crops purely as fodder for their livestock; some crop farmers grow fodder and sell it to pastoral farmers.

Pastoral farmers are also known as graziers and in some cases pastoralists. Pastoral farming is a non-nomadic form of pastoralism in which the livestock farmer has some form of ownership of the land used, giving the farmer more economic incentive to improve the land. Unlike other pastoral systems, pastoral farmers are sedentary and do not change locations in search for fresh resources. Rather, pastoral farmers adjust their pastures to fit the needs of their animals. Improvements include drainage (in wet regions), stock tanks (in dry regions), irrigation and sowing clover.

Pastoral farming is common in Argentina, Australia, Brazil, Great Britain, Ireland, New Zealand, and the Western United States and Canada, among other places.

Beef cattle reared in a pastoral farming manner

CHARACTERISTICS

There are many factors that are taken into account to decide what type of farming should take place on a certain area of land including, topography, altitude, exposure, and rainfall. Soil plays a large role in determining how land will be used. Mollisol lands are typically described as semi-arid to semi- humid areas that are grassy. This is where most intensive cattle operations occur which produce beef and dairy. Although a majority of pastoral farming is conducted in Mollisol lands, pastoral farming can also be found in areas with soil made up of Entisol, Aridisol or Alfisol. Aside from soil order, pastoral farming is more likely to be found than arable farming in areas with steep slopes, cold strong winds and a wet climate. All of these conditions are more advantageous to raising live stock than crops. Raising of sheep is often found in cooler regions with steep hills and above average rain fall. The wetness of the area and incline would make it unsuitable to grow crops. A similar conclusion is drawn by looking at dairy farms which are often found in warm wet climates.

LOCATIONS

Argentina

The first settlers of Argentina arrived approximately twelve thousand years ago and survived by hunting and gathering. During the 16th century, the Incan Empire dominated the area. The Incas were highly advanced for their time and were able to domesticate llamas and alpacas. in 1532, the Spaniards arrived and found open grasslands perfect for their cattle and horses to graze. Quickly, these herds grew and changed the environment making it more nutritious and fertile. The large cattle population was then hunted and used for economic prosperity. This marked the beginning of pastoral farming in Argentina as land began to be used for raising cattle. These farms became known as "estancias" meaning, "stay."The estancias were spread around 200 square kilometers could support about 20,000 cattle. and at the 19th century, sheep were added to the estancias. The Pampas saw a shocking growth in livestock population. The main animal products of the time became hides, fat, wool and salted meat.

Today, Argentina's livestock production is divided into two sectors- a modernized commercial part and a communal part. Pastoral farming still represents a large portion of the communal sector. Communal agriculture face disadvantages compared to their commercial counterparts as they have limited access to new technology and external inputs. In 2001, the country's stock included approximately 48 million cattle, 13.5 million sheep and 1.5 million horses. The number of sheep was reduced greatly recently because of a dramatic decrease in the price of wool. As of 2001, the majority of beef, lamb, and milk production in Argentina was domestically consumed.

Australia

Pastoral farming arrived in Australia in 1836 with the importation of sheep and cattle from New South Wales. Australia faces a tough climate with approximately 70% of its landmass being classified as arid or semi-arid. This made South Australia a perfect candidate for grazing since its climate was not suitable for arable farming such as wheat production. In the 1840s began to focus on wool production and prospered. Unfortunately in the 1860s South Australia faced serious droughts. To prevent future instances from occurring, the agricultural industry underwent serious specialization measures and focused on improving the pastures of pastoral farmers. The system they developed included low-density grazing of sheep and cattle. In addition, water was pumped from under ground sources by wind power. These improvements helped to better satisfy the livestock's needs.

Ireland

Bronze Age people first introduced pastoral farming to Ireland. The area of Burren was popular for settlers because of its dry and fertile soils. The first pastoral farmers were known for herding cattle, sheep and goats. Pastoral farming could also be found in the uplands such as Turlough Hill. Dating back to medieval times, farmers used the hill as a tool during winter months. Cattle would graze there because foliage was always available because the rock retained heat. Also, the lime rich soil would provide animals with calcium and other minerals to help increase their fat levels. In the early 19th century, sheep herding was most popular in the Burren.

By the 20th century however, there was a shift in importance from sheep to cattle. Trends in livestock units show that goats were just under 4% of cattle livestock units in the 1930s but had decreased to just over .5% by 1980. Today in Ireland, farm sizes have increased, the number of full-time farmers have decreased and heavier continental breeds have become more popular in comparison to the past.

New Zealand

New Zealand's pastoral sector is made up of cattle, deer and sheep. In the 1920s, meat, butter, cheese and wool, accounted for over 90% of the country's exports. The trend of high pastoral farming has continued to present day. The modernization of arable farming and horticulture have been met with equal advances in pastoral farming.

While sheep and beef farming use most of the land in New Zealand, the dairy industry is increasing in importance. The dairy sector began in 1814 when two cows and a bull were imported to New Zealand and the industry has been strong ever since. The New Zealand Institute of Economic Research (published Dec 2010) estimates the industry contributes around 2.8% to New Zealand's GDP and 10.4 billion of export earnings. Dairy production has risen 77 percent over the past 20 years - from three million dairy cattle in 1989 to six million dairy cattle in 2009.

Categories

Intensive Farming

Intensive farms generally take up a fairly small area of land, but aim to have a very high output, through massive inputs of capital and labour. These farms use machines and new technologies to become as efficient and cost-effective as possible, an example being the Concentrated Animal Feeding Operation.

Intensive agriculture can be seen in many places around the world, such as the Canterbury Plains of New Zealand, pig farming in Denmark and rice cultivation in the countries of South East Asia. All use technology appropriate to their country to enable them to get the highest yields from their land.It is labour intensive, capital intensive and machine intensive.

Extensive Farming

Extensive farming is the direct opposite of intensive farming. The farms are large in comparison to the money injected into them or the labour used. The cattle ranches of central Australia are a good example of extensive agriculture, where often only a few farm workers are responsible for thousands of acres of farmland.

Another example of extensive farming can be seen in the massive cattle ranches of Brazil. These involve clearing vast areas of rainforest (the trees are often burnt rather than chopped down and sold) to make way for the cattle ranch. The cattle quickly eat the remaining vegetation and begin to cause massive problems of soil erosion. Extensive farming is also the production of livestock and crops on large piece of land having small output in return. Less attention is given here as compared to intensive farming.

CONSTRAINTS

Livestock farming faces many potential problems and constraints. First, there is a high chance of exportation problems. With a high volume of trade there is also a high risk of spreading diseases from country to country. Britain saw the damage communicable animal diseases could cause in the 1980s and 1990s with the outbreak of mad cow disease. In this instance, the disease was able to infect humans as well. It is very important that when dealing with pastoral farming that the health of the animals is a high priority.

For low-income developing countries, heavily investing in pastoral farming is very dangerous because expected returns can decrease significantly due to unforeseeable events such as climate change or natural disasters. If the country did experience an unfortunate events, there would be no other major industry to stabilize the economy or other goods to use as alternatives. This is exemplified by the drought that Australia experienced in the 1860s which severely limited food sources.

Environmental degradation is another concern for livestock farmers. Environmental degradation occurs when the world's major resources are exhausted which causes undesirable effects on the environment. One major aspect of this degradation is the depletion of fresh water. Fresh water is needed to feed the livestock and to keep the animals in good health. Also, if fresh water were no longer available the moisture levels of soil would be greatly affected. This in turn would cause pastoral farmers to suffer considerably since the quality and existence of their pastures is based on having moist soil.

REFERENCES

1. "Distribution of Farming", British Broadcasting Corporation,
2. , Garbulsky & Deregibus "Country Pasture/ Forage Resource Profiles- Argentina", Food and Agriculture Organization of the United Nations,
3. http://www.southaustralianhistory.com.au/overview.htm
4. , Jayasuriya "Farming Systems in Australia", Food and Agriculture Organization of the United Nations,
5. " Recognising European pastoral farming systems and understanding their ecology ", European Forum on Nature Conservation and Pastoralism, June 2000
6. Bohnsack U. & Carrucan P. 1999. An assessment of farming prescriptions under

the Rural Environment Protection Scheme in the uplands of the Burren karstic region, Co. Clare. The Heritage Council, Kilkenny.

7. Ross Galbreath (1998). DSIR: Making Science Work for New Zealand : Themes from the History of the Department of Scientific and Industrial Research, 1926-1992. Victoria University Press. p. 58.
8. "Agriculture- Pastoral", Ministry for Primary Industries, October 2013
9. "Protecting animal health and preventing disease, including trade", Department for Environment, Food & Rural Affairs, March 2014
10. "Controlling animal disease", Department for Environment, Food & Rural Affairs, March 2014
11. Mahmud & Randy, "Risk aversion in low income countries", International Food Policy Research Institute, 2007
12. Ragab, Ragab, and Christel Prudhomme. "Soil and Water: Climate Change and Water Resources Management in Arid and Semi-Arid Regions: Prospective Challenges for the 21st Century." Biosystems Engineering 81.1 (2002): p 3-34. This in turn would cause pastoral farmers to suffer considerably since the quality and existence of their pastures is based on having moist soil.

CHAPTER 5

Contour Plowing at Hills

Chapter Contents

- **Introduction**
- **Modern History**
- **Contour Trenching**

Contour plowing is a type of plowing that follows the curves of the landscape instead of plowing straight rows up and down hills and valleys. Contour plowing creates ridges that conserve rainwater and prevent the removal of topsoil from the area in the event of heavy rains.

INTRODUCTION

Contour plowing or contour farming or Contour bunding is the farming practice of plowing and/or planting across a slope following its elevation contour lines. These contour lines create a water break which reduces the formation of rills and gullies during times of heavy water run-off; which is a major cause of soil erosion. The water break also allows more time for the water to settle into the soil. In contour plowing, the ruts made by the plow run perpendicular rather than parallel to slopes, generally resulting in furrows that curve around the land and are level. This method is also known for preventing tillage erosion. Tillage erosion is the soil movement and erosion by tilling a given plot of land. A similar practice is contour bunding where stones are placed around the contours of slopes.

Soil erosion prevention practices such as this can drastically decrease negative affects associated with soil erosion such as reduced crop productivity, worsened water quality, lower effective reservoir water levels, flooding, and habitat destruction. Contour farming is considered an active form of sustainable agriculture.

Contour plowing, Pennsylvania, 1938

"Contour bunding", Catalonia, 2007

The Phoenicians first developed the practice of contour farming and spread it throughout the Mediterranean. However, the Romans preferred cultivation in straight furrows and this practice became standard.

In recent years, contour farming systems are considered as field and farm practices to build climate-smart agriculture as they improve water-use efficiency and conservation.

MODERN HISTORY

This was one of the main procedures promoted by the US Soil Conservation Service (the current Natural Resources Conservation Service) during the 1930s. The US Department of Agriculture established the Soil Conservation Service in 1935 during the Dust Bowl when it became apparent that soil erosion was a huge problem along with desertification.

The extent of the problem was such that the 1934 "Yearbook of Agriculture" noted that *Approximately 35 million acres [142,000 km²] of formerly cultivated land have essentially been destroyed for crop production. . . . 100 million acres [405,000 km²] now in crops have lost all or most of the topsoil; 125 million acres [506,000 km²] of land now in crops are rapidly losing topsoil.* This can lead to large scale desertification which can permanently transform a formerly productive landscape to an arid one that becomes increasingly intensive and expensive to farm.

The Soil Conservation Service worked with state governments and universities with established agriculture programs such as the University of Nebraska to promote the method to farmers. By 1938, the introduction of new agricultural techniques such as contour plowing had reduced the loss of soil by 65% despite the continuation of the drought.

Demonstrations showed that contour farming, under ideal conditions, will increase yields of row crops by up to 50%, with increases of between 5 and 10% being common. Importantly, the technique also significantly reduces soil erosion, fertilizer loss, and overall makes farming less energy and resource intensive under most circumstances. Reducing fertilizer loss not only saves the farmer time and money, but it also decreases risk of harming regional freshwater systems. Soil erosion caused from heavy rain can encourage the development of rills and gullies which carry excess nutrients into freshwater systems through the process of eutrophication

Contour plowing is also promoted in countries with similar rainfall patterns to the United States such as western Canada and Australia.

The practice is effective only on slopes with between 2% and 10% gradient and when rainfall does not exceed a certain amount within a certain period. On steeper slopes and areas with greater rainfall, a procedure known as strip cropping is used with contour farming to provide additional protection. Contour farming is most effective when used with other soil conservation methods like strip cropping, terrace (agriculture) farming, and the use of a cover crop. The proper combination of such farming methods can be determined by various climatic and soil conditions of that given area. Farming sites are often classified into five levels: insensitive, mild, moderate, high and extreme, depending on the regions soil sensitivity.

P. A. Yeomans› Keyline Design system is critical of traditional contour plowing techniques, and improves the system through observing normal land form and topography. At one end of a contour the slope of the land will always be steeper than at the other. Thus when plowing parallel runs paralleling any contour the plow furrows soon deviate from a true contour. Rain water in these furrows will thus flow sideways along the falling «contour» line. This can often concentrate water in a ways that exacerbates erosion instead of reducing it. Yeomans was the first to appreciate the significance of this phenomenon. Keyline cultivation utilizes this «off contour» drift in cultivating furrows to control the movement of rain water for the benefit of the land. (Priority One History of Twentieth Century Soil Conservation and Keyline.)

Contour bunding has been widely adopted in Burkina Faso after it was suggested by British Oxfam worker Bill Hereford in the beginning of the 1980s.

Its simple definition is that it is a process of parallel ploughing to the contours of hill slopes to form a natural barrier for water to flow down the slope.

CONTOUR TRENCHING

Contour trenching is an agricultural technique that can be easily applied in arid sub-Sahara areas to allow for water, and soil conservation, and to increase agricultural production.

A deep continuous contour trench in Maharashtra, India

A contour trenchline in India

Technique

Trenches can be artificially dug along the contour lines. Water flowing down the hill is retained by the trench, and is infiltrating the soil below.

Between two trenches crops can benefit during the growing season (when there is less rain) from the subsoil water reserve gathered during the rainy season.

Advantages

Immediate advantages are the following:

- The rain water does not immediately run off the hill,
- Water does not evaporate uselessly
- The water balance is enhanced
- Crops do not suffer later on from water shortage,
- Fertile soil particles are not lost by water and wind erosion.
- When the sun shines on the water, light and heat are reflected onto plants on the northern shore of the trench, this effect and the increased humidity create micro climates in the area. These micro climates can support plants from different hardiness zones.

Implementation

Depending on the slope of the hill, the parallel trenches can be closer or further from one another.

Manually dug trenches are smaller. Machine dug trenches can be deeper. The dimensions, and the format of the trench should correspond to the local climate and soil conditions. The trench should be big enough to keep all the water; no water should spoil over the downhill border. The upside of the trench should be protected against erosion, by means of e.g. grass, shrubs, or fabric.

REFERENCES

1. "Contour Farming for Cropland in the Pacific." University of Hawai'i - College of Tropical Agriculture and Human Resources. Ed. University of Hawaii at Manoa. USDA NRCS Practice (330), Web. <http://www.ctahr.hawaii.edu/wq/publications/Final_FactSheets/ContourFarmingCrop330.pdf>.
2. Van Oost, K. K.; Govers, G. G.; de Alba, S. S.; Quine, T. A. (2006). "Tillage erosion: a review of controlling factors and implications for soil quality". Progress In Physical Geography. 30 (4): 443–466. doi:10.1191/0309133306pp487ra.
3. Penn State College of Agricultural Sciences. "Tillage Erosion." Agronomy Guide (Penn State Extension). Penn State College of Agricultural Sciences, 2013. <http://extension.psu.edu/agronomy-guide/cm/sec1/sec11e>.
4. Xu, Lifen; Xu, Xuegong; Meng, Xiangwei (2013). "Risk assessment of soil erosion in different rainfall scenarios by RUSLE model coupled with Information Diffusion Model: A case study of Bohai Rim, China". CATENA. 100: 74–82. doi:10.1016/j.catena.2012.08.012.
5. Roychoowdhury, Banerjee U; Sofkova, S; Yah, J (2013). "ORGANIC FARMING FOR CROPIMPROVEMENT AND SUSTAINABLE AGRICULTURE IN THE ERA OF CLIMATE CHANGE". Online Journal of Biological Sciences. 13 (2): 50–65.
6. Hogan, Michael C., and GreenFacts. "Desertification." Encyclopedia of Earth., 22 July 2010. Web. <http://www.eoearth.org/view/article/151708/>.
7. "Contour Farming." Encyclopædia Britannica. Ed. Encyclopædia Britannica., 2013. Web. <http://www.britannica.com/EBchecked/topic/135192/contour-farming>.
8. Hasholt, Bent; et al. (1997). "Sediment delivery to streams from adjacent slopes on agricultural land in Denmark". IAHS Publications-Series of Proceedings and Reports-Intern Assoc Hydrological Sciences. 245: 101–110.
9. NRCS Conservation Practice Standard 330-Contour Farming
10. Reinhardt, Claudia, and Bill Ganzel. "Contour Plowing & Terraces during the 1930s Depression." Living History Farm., 2003. Web. <http://www.livinghistoryfarm.org/farminginthe30s/crops_11.html>.
11. Zhang, Ronghua; Liu, Xia; Heathman, Gary C.; Yao, Xiaoyou; Hu, Xuli; Zhang, Guangcan (2013). "Assessment of soil erosion sensitivity and analysis of sensitivity factors in the Tongbai–Dabie mountainous area of China". CATENA. 101: 92–9. doi:10.1016/j.catena.2012.10.008.
12. http://www.fws.gov/fire/ifcc/esr/Treatments/contour-trenching.htm

13. http://www.academia.edu/8072565/Decision_support_system_for_contour_trenching
14. http://www.indiaenvironmentportal.org.in/files/Watershed%20development.pdf
15. http://www.fs.fed.us/rm/pubs_int/int_gtr334.pdf
16. Just diggit
17. Documentary on Peter Westerveld and the Naga Foundation
18. Westerveld Conservation Trust

CHAPTER 6

Soil Conservation Techniques for Hillside Farms

Chapter Contents

- **Introduction**
- **Traditional Honduran Hill Side Farming Techniques and Resulting Problems**
- **Contour Barriers (Live, Dead and Mixed Barriers)**
- **Soil Fertility and its Maintenance**
- **Extension Methodology**

INTRODUCTION

Agricultural extentionists in a mountainous country, such as Honduras, are faced from the start with an almost infinite number of variables which may affect decisions about optimal land use. The rugged topography provides a wide variety of macro and microclimates, bedrock types, soil types, soil depths, and drainage conditions. Equally important at times may be the variation in economic resources and markets available to farmers. Therefore, as an extensionist strives to develop solutions to the problems most seriously affecting farmers of an area, it is found that each plot must be examined individually in light of the physical, economic, and possibly social or political factors which may affect an individual farmers land use decisions. The purpose of this guide is to provide agricultural extensionists with basic information which will help them design plans for the conservation of soils and the management of water runoff in specific agricultural plots. It was written based on experiences with small hillside farms in Honduras, and takes into account the resources and constraints commonly encountered there. Included in this guide are also some demonstrations which may prove helpful in the promotion of this technology. Through the use of this guide, it is hoped that extentionists will find it easier to solve some of the common problems faced by hillside farmers. This guide is not intended to replace more detailed technical works such as Hudson (1981) or Suarez Castro (1980). The main purpose of this guide is to provide a more concise reference work geared more specifically towards small hillside plots. In doing so, much technical information and many techniques appropriate to larger, more mechanized farms are committed. Information presented in other sources should be consulted, when possible, for the solution of specific technical problems. There are also many important topics not covered in this guide, such as improved seed varieties, pest control, marketing, etc., Which nevertheless should be considered by extensionists in developing a complete plan for improving farming practices of an area.

TRADITIONAL HONDURAN HILL SIDE FARMING TECHNIQUES AND RESULTING PROBLEMS

Traditional hillside land-use patterns in much of Honduras are based roughly cr. the following series of events:

A forested lot is clearcut and residues are burned

Crops are planted with rows oriented up and down the hillside

Crop yields decline each year as soil erodes

The land is planted in perennial pasture grasses or abandoned after having been cultivated for a relatively short time (often only 1-20 years, depending on site-specific factors).

This pattern of land use may avoid some problems with crop pests and the need for supplemental fertilizer. However, in a densely settled area land soon becomes scarce, and mature-forested, fertile plots are not always available to be cleared and cultivated. In such densely settled areas, a land use pattern which results in a permanent cultivation of the same plots is desired. There are three related problems associated with this traditional land use patterns which must be corrected if permanent cultivation practices are to be successful!: soil erosion, rapid water runoff, and decreased soil fertility. As the native vegetation is cut and burned, the soil surface is exposed to the impact of raindrops. The force of the raindrop impact dislodges soil particles. These soil particles and the valuable nutrients they contain can then be carried out of the area by the water flowing over the soil surface. This rapid rainfall runoff means that less water infiltrates into the soil to be available later, exagerating any natural flood/ drought cycles. When the native vegetation is intact, the force of falling raindrops is mostly absorbed by the vegetation, as there is much less bare ground surface exposed. The decaying leaf litter on the ground and in the upper soil layers also protects the soil, acting as a sponge, absorbing much of the rainfall and decreasing the amount that can freely run over the land surface. Since more of the rainfall remains in the soil rather than running off, the soil retains its moisture . longer in the absence of rainfall. Since fewer soil nutrients are lost in runoff water, soil fertility is maintained. A comparison of these two scenarios illustrates how in the traditionally farmed field, characterized by the absence of protecting vegetation, the rainfall impact and the loss of soil particles, dissolved nutrients, and water are all increased.

Soil Conservation Strategies

There are two types of strategies which may be followed in attempts to reduce the detrimental environmental effects of the agricultural activity. The most effective is to avoid the damage due to rainfall impact by minimizing soil disturbance and promoting practices which maintain a ground cover. The second, less effective, but easier to integrate into traditional land use practices, is to continue many of the typical agricultural techniques which result in rainfall impact, but minimizing soil loss and water runoff by crop rotations or by the placement of structures (barriers, ditches, terraces) to reduce the movement of soil and water along the soil surface. These two types of strategies make up the basis of all the conservation schemes discussed here. They are discussed as separate techniques, but the best control of soil erosion, water runoff, and maintenance of soil fertility results from combining all of the complementary techniques appropriate for a particular cultivation system.

Strategies Aimed at Minimizing Soil Disturbance

1. Protecting Native Vegetation

Ideally, the environmental damages associated with agricultural activities could be avoided by protecting the native vegetation, which keeps the problem from arising. This method, however, ignores man's dependence on traditional agricultural techniques to support himself and is difficult to promote, especially as human populations increase. This strategy does however, provide other valuable benefits such as maintaining a clean, reliable supply of drinking water, and reserves for native fauna and flora, which also become increasingly important as the human population increases.

2. Replanting Native Vegetation

In view or the benefits of native vegetation, replanting of previously cleared land can be an effective method for protecting land, water supplies, and native wildlife. Reforestation (or replanting of grasses or shrubs, depending on natural vegetation type) can be carried out using native species or introduced species adapted to the local conditions, having some desirable characteristic (fast growth rate, the ability to fix nitrogen, forage or wood value, etc), and which will facilitate the return of a protective vegetative canopy A description of some species is given in Firewood Crops Shrub and Tree Species for Energy Production (N.A.S., 1983).

Reforestation

3. Perennial Crop Cultivation Systems

Another technique which results in a minimized soil disturbance is the planting of perennial crops, such as fruit trees or pasture grasses, rather than annual crops such as corn or beans. In this manner, after the initial disruptive clearing and planting of the land a permanent ground cover is attained and the environmental damages associated with the raindrop impact are lessened.

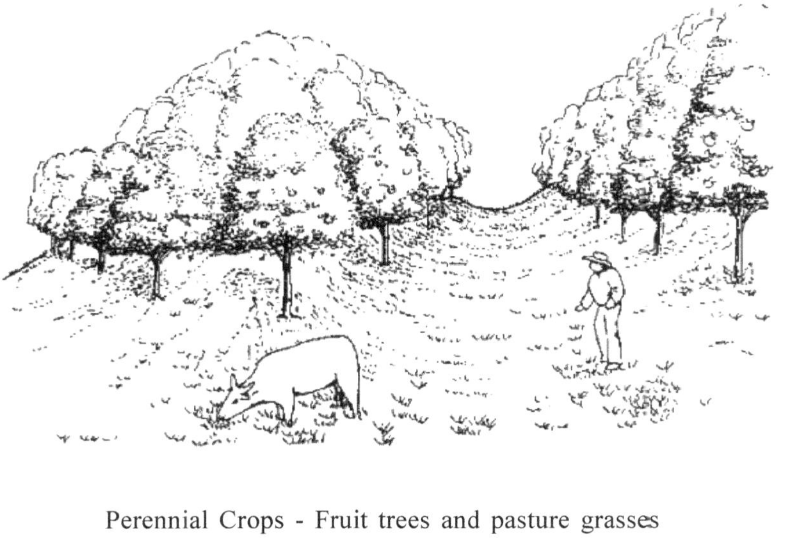

Perennial Crops - Fruit trees and pasture grasses

4. Use of Ground Cover while Cultivating Annual Crops

There are techniques for maintaining a protective ground cover even while cultivating annual crops. These include minimum tillage and mulching systems. In minimum tillage systems, only the narrow row where the seeds are planted is tilled and the remaining ground surface is left intact. Because these intact areas are more compacted and have a law weed covering, they are more erosion resistant. Mulching is the use of dead material or the planting of a ground cover which results in a covering of the bare soil areas in a field. As in the other techniques, the covered soil is less susceptible to erosion than bare soil.

Minimum tillage cultivation system

CONSERVATION STRATEGIES

Aimed at minimizing soil loss and increasing water infiltration in cultivation systems where there is extensive soil disturbance

Even though the techniques described in the previous Chapter are more effective in reducing the environmental damages associated with man's agricultural activity, the techniques described in this Chapter are much easier to promote because they are much easier to integrate into the traditional cultivation systems.

Crop Rotation

The use of a crop rotation sequence, rather than continual successive plantings of the same crop can play a role in reducing soil erosion as well as provide other benefits (vary the rooting depth and thereby moisture and nutrient uptake, restore soil structure, break pest and disease cycles, help maintain soil fertility, Hudson 1981). When land is used repeatedly for cultivating relatively open row crops (i.e. corn, beans, tobacco, etc.), a rotation with a denser green manure or forage crop can reduce erosion because of the ground cover provided. Also the maintenance of soil fertility and structure will allow the next planting of the more open crop to grow more vigorously and develop better root systems, thereby helping reduce some soil loss. Generally, however, on sloping lands (5-10% +), crop rotations alone will have little effect on erosion, and should be used in combination with other techniques.

CONTOUR BARRIERS (LIVE, DEAD AND MIXED BARRIERS)

Contour barriers are contour strips which intercept downslope flowing water and soil particles. These barriers slow down the water movement and reduce its erosive force. They also filter out and trap many of the suspended soil particles, keeping them from being washed out of the field. A long term advantage of barriers is that soil tends to build up behind them, creating a terrace effect. Barriers can be classified as live (strips of living plants), dead (rocks, crop residues), or mixed (a combination of the previous two). Live barriers are strips of vegetation planted along the contour which serve to anchor the soil in place with their roots and to slow down the movement of water downslope with their stems and leaves. They are planted above hillside ditches to prevent them from filling with soil and also by themselves to prevent hillside erosion. me most common types of live barriers are plants of the grass family because of the dense foliage and root nets produced. As soil builds up behind them, a bench terrace with a grass protected riser (sloping bank) is formed (see Chapter on bench terraces). In addition the grasses are valuable as forage for animals, or for human consumption in the case of sugar cane or lemmon grass. Many species of plants have great potential as useful live barriers, some possibilities are listed in Table. The table is by no means all inclusive, the possibilities are almost limitless, especially when considering agroforestry systems where contour strips of nitrogen-fixing and/or wood or fruit producing trees may be used as barriers. An effort should be made to discuss the available barrier plants in the area with farmers so that each one can select the ones most suited to their needs. As a management practice it is advisable to cut forage barrier plants and carry them to livestock rather than letting livestock graze them. If the barriers are not well established, the animals may uproot or overgraze them, thus wiping out the barrier. If farmers use the practice of opening their fields to grazing after the harvest, then an unpalatable barrier or planting early in the year to ensure a well established barrier is necessary. Maintenance of the live barriers is extremely important. Open spaces should be replanted so that the barrier forms an effective soil filter. Barriers should be cut or pruned to avoid excessive shading or root competition with the crops.

Some suggested live barrier plants

COMMON NAME	SCIENTIFIC NAME	CHARACTERISTICS
Sugar Cane (Caña de Azucar)	Saccharum officinarum	Establishes vegetatively from rhizome or stems, forms thick barrier, may also serve as windbreak, useful for forage or human consumption, may cause excessive shading
Guatemala Grass (Zacate o Caña Guatemala)	Tripsacum laxum	Establishes vegetatively from rhizome or stems, excellent forage, susceptible to overgrazing smaller plant than sugar cane
King Grass (Zacate Rey, Taiwan o Elefante)	Pennisetum purpureum	Establishes vegetatively from rhizome or stems, excellent forage grass, tall plants, may cause excessive shading or competition if not cut
Merker Grass (Merkeron, Napier)	Pennisetum purpureum	Establishes vegetatively from rhizome or stems, smaller and less succulent than King Grass, may be problem weed spreading vegetatively and by seed if not cut
Guinea Grass (Zacate Guinea)	Panicum maximum	Establishes easily from rootstock or seed, excellent forage grass, relatively short, forms thick bunches
Bermuda Grass, Star Grass (Zacate Bermuda, Estrella)	Cynodon spp.	Establishes by stolons, may be a problem weed because of its vigorous spreading, suitable for stabilizing drainages, not for contour barriers, good forage
Lemmon Grass (Zacate Limón)	Andropogon citratus	Establishes well from rootstock, unpalatable to cattle, used in making tea, forms short, thick bunches
Maicillo	Sorghum sp.	Establishes by seed, if planted closely forms excellent barrier, good forage, produces grain, grows well in arid zones, tall plant and may cause excessive shading, short-lived (2-3 years), may be alternate host for certain corn diseases
Jaragúa	Hyparrhenia rufa	Establishes by seed, forms thick closely spaced bunches, grows well in arid zones, good forage, problem weed if not cut before flowering
Valeriana	Vetiveria sp.	Establishes from rootstock, forms low thick bunches unpalatable to cattle, medicinal value
Cardamine (Cardamomo)	Elettaria cardamomum	Establishes as vegetative bulb or transplanted from seedbed, forms thick bunches up to 3m tall, valuable seed crop. shade tolerant, only in cool moist climates, may cause excessive shading or competition
Gladiola[1] (Varsovia)	Gladiolus sp.	Establishes easily from bulbs, value as ornamental
Pineapple[1] (Pine)	Ananas comosus	Establishes easily from vegetative buds, grows well in poor sandy soils, edible fruits
Motate[1] (Piñuela)	Bromelia sp.	Establishes easily from vegetative buds, grows well in wide variety of climates and soils, edible fruits
Leucaena[2]	Leucaena spp.	Establishes from seeds, leaves useful as nitrogen-containing forage or mulch, grows rapidly, produces firewood, best at law elevations (less than 1000m)
Madriago[2] (Madre de Cacao)	Gliricidia septum	Establishes from branches planted in post farm, also from seeds, leaves useful as nitrogen-captaining forage or mulch, produces firewood
Yucca[2] (Joshua tree, izote)	Yucca elephantipes	Establishes from stakes, edible flowers, used in live fences or as an ornamental

(1) These plants must be planted closely spaced in 2 or 3 rows in order to form effective barriers.

(2) These plants are most effective as barriers when the rows are reinforced with crop residues or rocks to form a mixed barrier.

Contour live barriers

In planting contour barriers, the spacing from one barrier to the next depends on the slope of the land, with barriers spaced closer together on steep slopes and farther apart on gentler slopes.

Spacing of Contour Barriers According to the Slope

SLOPE (%)	ANNUAL CROPS Distance (m)	PERRENIAL CROPS Distance (m)
5	20.0	25.0
10	15.1	20.1
15	10.1	18.2
20	9.2	15.3
25	8.2	15.5
30	6.8	12.5
35	6.4	12.7
40	6.5	9.7
60		7.0

From Suarez Castro 1980, modified to show distance along ground surface rather than horizontal distance. Contour planting of a crop and contour strip cropping are also techniques which may be considered functionally as live barriers since they are contour plantings which serve to control hillside erosion. The planting of windbreaks is also a use of live barriers (trees or tall grasses) in this case to avoid erosion or crop damage due to wind rather than water. Dead barriers function similar to live barriers, the difference being that they are composed of rocks, plant residues, or other non-living materials. If rocks are present in a field, it is useful to construct these and in the process make the soil easier to work. Another advantage is that they can be completed during the dry season, meaning that they are in place and functioning at the start of the rainy season. If enough rocks are present' the barriers can be constructed as rock walls of sufficient height so that bench terraces are famed as the soil fills in behind each wall. If sufficient rocks are not present, the barriers lose their effectiveness as the soil fills in behind them, and they should be supplemented with the planting of live barriers.

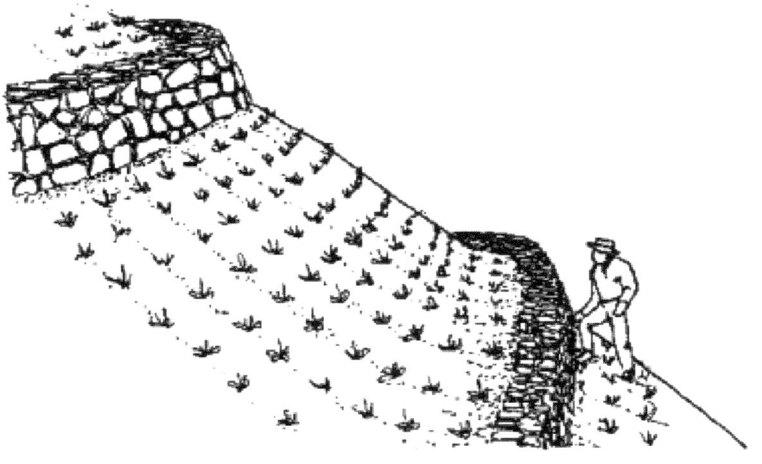

Contour dead barriers

The construction of dead barriers is simple, but requires a lot of manual labor. Once contour lines are marked out according to the spacings giving in Table,a hoe is used to form a furrow which serves to anchor the barrier (~ 20 cm deep). men the materials are laid out to form walls along the contour lines. In some cases, mixed barriers are used, a combination of live and non-living materials. This can consist of strips of trees with the intervening spaces filled with rocks or crop residues., or combinations of grass and rock barriers. If using plant residues, be aware that as they decompose they lose their effectiveness as barriers and will erode away, needing to be supplemented. Contour furrows or raised planting beds may also be considered functionally as contour barriers. These techniques are useful in the cultivation of vegetables, basic grains, or any raw crops in which the soil is to be intensively cultivated. The contour furrows or beds serve various purposes: reducing water movement and therefore soil erosion down the slope, permitting drainage of excess soil moisture from the planting bed, providing for a more even distribution of irrigation or rainwater, and avoiding compaction of the planting bed surface by providing walkways The construction is simple. The furrows are plowed if possible, along measured contour lines, then cleaned out with a hoe or shovel, spreading the soil in the space between the furrows. The width between furrows is variable and depends on the crop and any irrigation requirements.

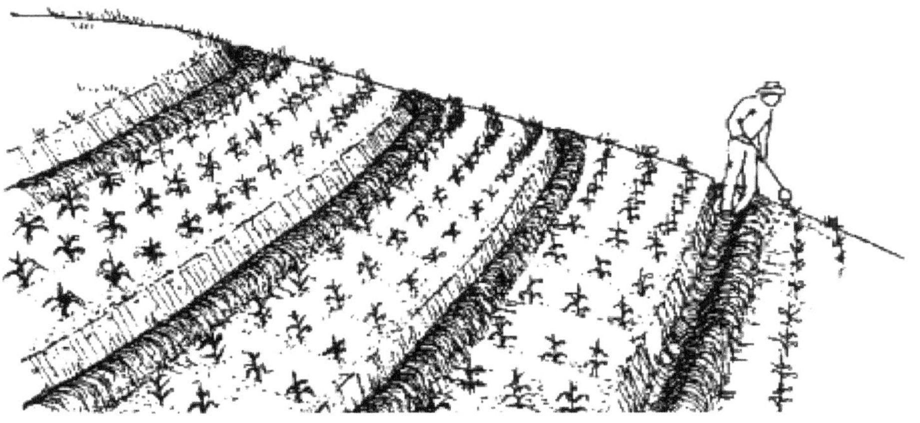

Contour planting beds

Contour ditches serve many of the same purposes as contour barriers, in addition they completely stop dawnslope water movement as the water falls into the ditch. These structures are some of the mast useful for small-scale hillside farming singe they require less work than terraces, are simple to build, and can be used to either divert or to retain water. If constructed at a 1% slope* they divert excess water to protected drainageways, reducing soil erosion and leaching of nutrients. The uppermost ditch, called stormwater drain by Hudson 1981, is very important if a great deal of water enters from above the field.

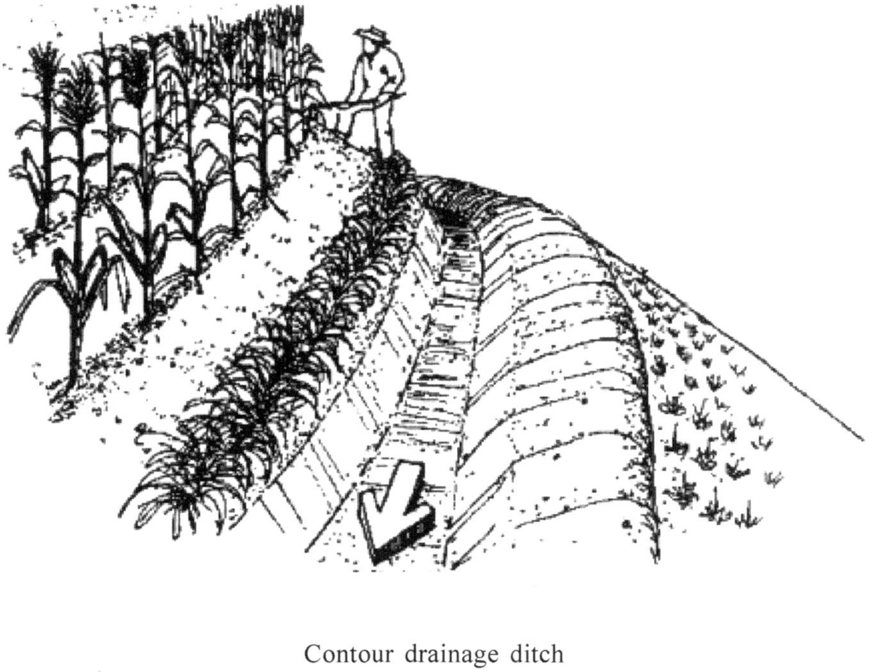

Contour drainage ditch

If it is desirable to retain as much water as possible, earthen dikes can be left in the ditches or the ditches can be constructed at a 0% slope so that water infiltrates into the soil and is not diverted outside the field.

Contour infiltration ditch

The ditches are constructed using pickaxes and shovels. If possible, plowing the contour line makes the construction much easier. The first stage in the construction is to excavate a 12 inch wide by 12 inch deep ditch (ditches can be constructed of any size, if desired). Then the banks are formed by cutting a slanted wall at each side. The removed earth is placed in a mound 6 - 9 inches below the lower lip of the ditch. A live barrier is necessary above the upper edge to prevent filling with soil. Especially on steeper slopes, it is often advantageous to plant the live barrier first, several months or one season in advance, so that the ditch will be adequately protected once built. Table should be consulted for the appropriate distance between ditches on hillsides of a given slope. If the ditches are to be dug with a 1% slope to drain excess water, the 1 or 2 meters before emptying into a protected drainageway should have a slightly steeper slope, (1-2%) to facilitate drainage. Care should also be taken not to drain excess water into neighboring fields, houses, or other areas which may result in problems. (See "Waterways for draining excess water from fields").

Spacing of contour hillside ditches

From: Suarez Castro 1980 (modified to show distance along ground surface rather than horizontal distance) Terraces (individual, discontinuous narrow, and continuous bench terraces) Terraces serve the functions of stopping downslope soil and water movement and also give the advantage of providing a flat surface for the planting of crops, thereby further reducing the possibility of erosion. Some important points should be noted before deciding to build terraces. Since the formation of terraces requires a maximum of soil disturbance and rearrangement, they are extremely susceptible to erosion if not properly measured, compacted, and maintained with risers(sloping banks) protected by vegetation. Also when considering terracing a steep slope (25-30% +) it should be noted that a much deeper cut and fill will be needed, much more land space will be lost in the sloping terrace walls, and the useful planting space is likely to be extremely narrow unless a very deep soil is present. Individual terraces are constructed to provide a level platform for the growth of an individual tree. They are always used in combination with another type of conservation structure, such as contour ditches, since the small platforms by themselves do not appreciably control surface water movement. When planting a hillside with individual terraces, the layout should be in an equilateral triangular or hexagonal pattern, with the distances varying according to the variety of tree being planted. The terraces generally have a diameter of 1 to 1.5 meters and should be throughly compacted to prevent collapsing.

Slope (%)	Annual crop		Perennial crop or pasture	
	Distance (m)	Maximum Length (m)	Distance (m)	Maximum Length (m)
2	42.0	90		
4	25.0	120		
6	19.3	160		
8	16.6	200		
10	14.9	260	40.2	140
12	13.8	280	33.5	140
14	13.0	300	28.9	140
16	11.4	340	25.3	160
18	10.2	380	25.0	180
20	9.2	420	24.0	200
22	8.4	470	23.2	200
24	7.7	500	21.4	210
26	7.2	500	19.8	220
28	6.6	500	19.5	220
30	6.3	500	18.9	220
32			18.7	220
34			18.6	230
36			17.7	230
38			16.9	230
40			16.2	230

Individual terraces.

Discontinuous narrow terraces (orchard terraces) provide both -a flat platform for planting crops and an inverse slope which allows it to serve as a drainage or infiltration ditch.

Discontinuous narrow terrace (orchard terrace)

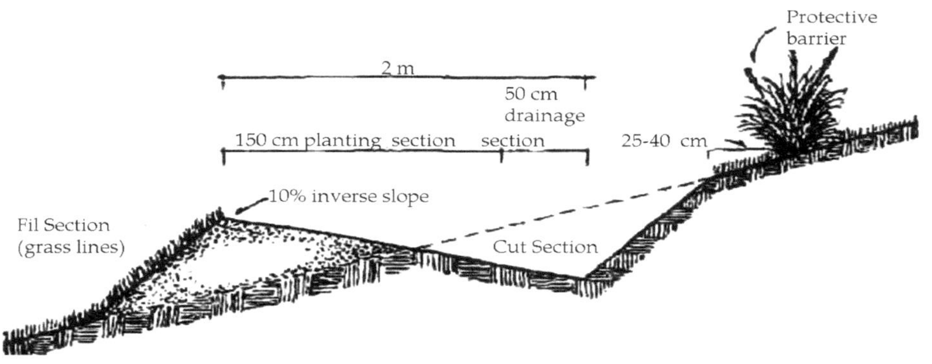

Discontinuous narrow terrace (X.S.)

Construction is most rapid in soft, deep soils. In these cases the earth from the cut Chapter is merely moved to the fill Chapter, compacted, and raked smooth. If it is judged necessary to remove the topsoil and redistribute it later, then a construction sequence such as sequence "B" given for the bench terraces is recommended. In order to keep the risers from eroding, grass should be planted on them. The drainage way for this type of structure should be constructed similar to that of the hillside ditch, slightly steeper right before emptying into the protected drainage area. (See 'Waterways for draining excess water from fields").

Discontinuous Narrow Terrace Construction Guide

slope (%)	DISTANCE BETWEEN CANALS (meters)	TOTAL CANAL WIDTH(Platform Plus Walls) (meters)
5	18	2.22
10	14	2.32
15	13	2.40
20	12	2.52
30	12	2.71
40	12	3.00

Continuous bench terraces are staircase-like structures which diminish erosion because the reshaping of the Land surface results in the planting of all crops on gently inversely sloping platforms. The mast feasible application is in intensively worked vegetable plots where each planting bed may be a separate terrace. Bench terraces can be designed level (0% slope) for water retention or with a slight slope (0-1%) to facilitate water drainage or distribution of irrigation water. (See "Waterways for Draining Excess Water From Fields']

Continuous bench terraces (X.S.)

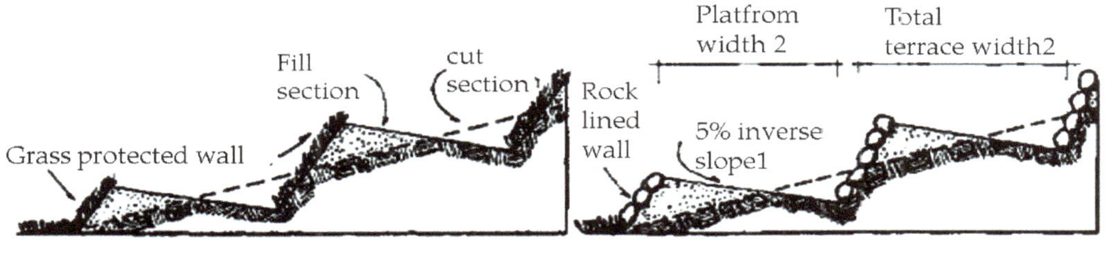

There are many different terrace design specifications published in soil conservation books, but only the basic features are described here. Table may be used as a general guide for designing terraces with the total terrace width dependent upon the slope of the hillside and the depth of fertile or tillable soil.

Bench Terrace Construction Guide

SLOPE (%)	SOIL DEPTH* (Meters)	TOTAL TERRACE WIDTH (Meters)	PLATFORM WIDTH (Meters)
	.2	1.68	1.30
	.3	2.5	1.94
20	.4	3.34	2.60
	.5	4.26	3.30

	.6	5.02	3.90
	.2	1.16	.80
	.3	1.72	1.20
30	.4	2.3	1.60
	.5	2.98	2.10
	.6	3.46	2.40
	.2	.90	.56
	.3	1.32	.82
40	.4	1.78	1.10
	.5	2.25	1.40
	.6	2.68	1.66
	.2	.74	.40
	.3	1.10	.60
50	.4	1.46	.80
	.5	1.84	1.00
	.6	2.20	1.20

Bench terraces are constructed using pickaxes, hoes, and rakes. In some areas, large, specially designed hoes are available. Construction is much easier if the Chapter of earth to be removed (cut Chapter) is plowed beforehand to loosen the soil. If construction is being undertaken during the rainy season, it is advisable to begin construction near the drainage area and with the uppermost terrace. In this way, any rainwater will drain off without damaging the terraces. There are several ideas as to the best method of constructing a series of terraces. One, which may require several years for completion, is the planting of a live barrier grass or the construction of rock walls along the contour. Over time terraces are formed as soil fills in behind them. Two other construction sequences are presented here.

1. The lower most terrace is formed first and compacted thoroughly

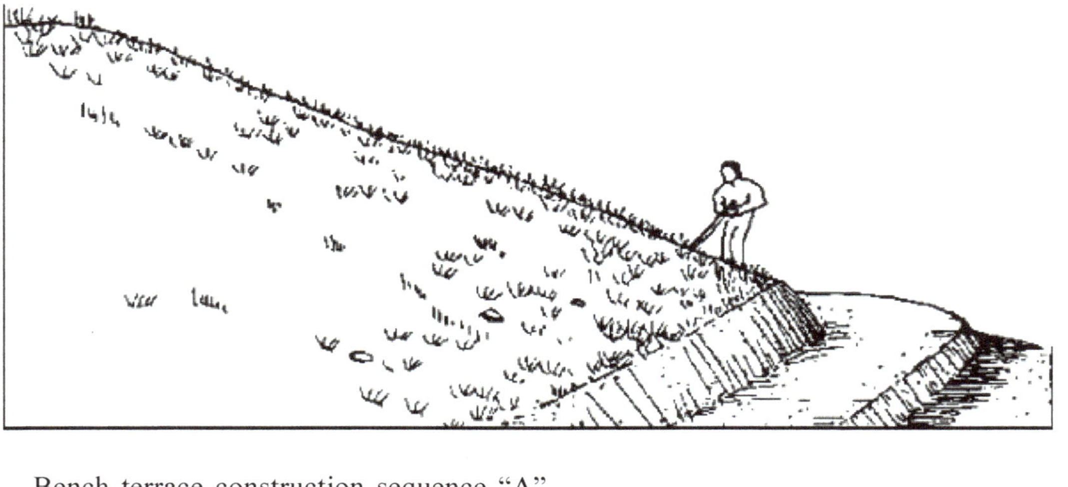

Bench terrace construction sequence "A"

2. The topsoil from the area of the next higher terrace is removed and distributed evenly over the lower terrace

3. The second terrace is formed and compacted, then covered with topsoil from the area of the third terrace.

4. Work progresses up slope, each newly formed and compacted terrace is covered with topsoil taken from the slope immediately above, grass is planted along the risersof all terraces.

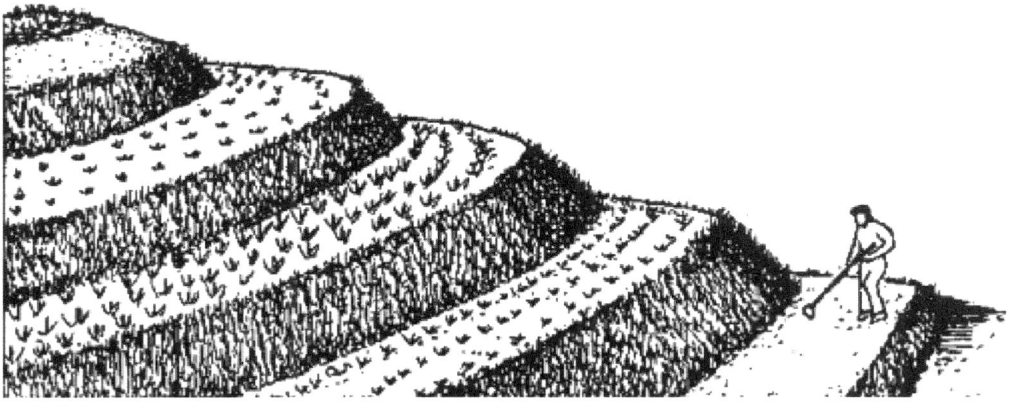

1. Terrace construction begins with the uppermost terrace and width the 2 meter segment nearest the drainage side. The topsoil is pulled over to one side of the Chapter.

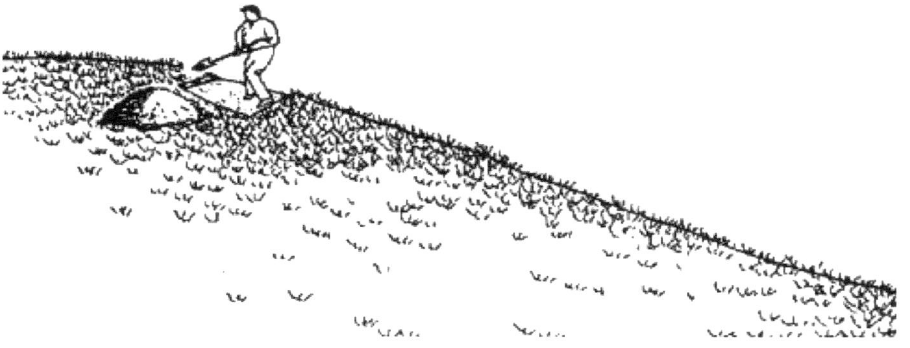

Bench terrace construction sequence ''B''

2. A well compacted Chapter of the terrace is formed

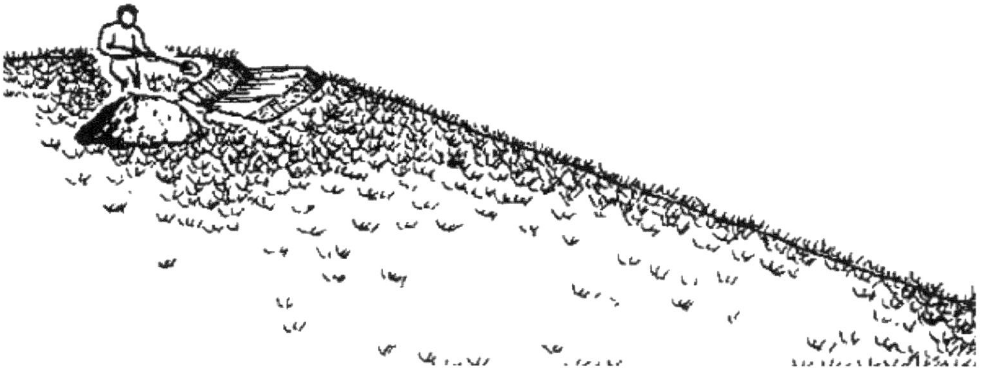

3. The topsoil is then redistributed over the same 2 meter terrace Chapter.

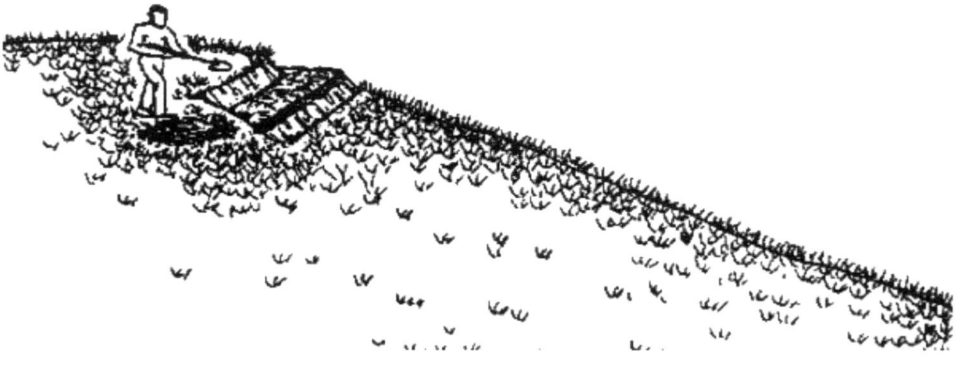

4. Work progresses sideways along the uppermost terrace.

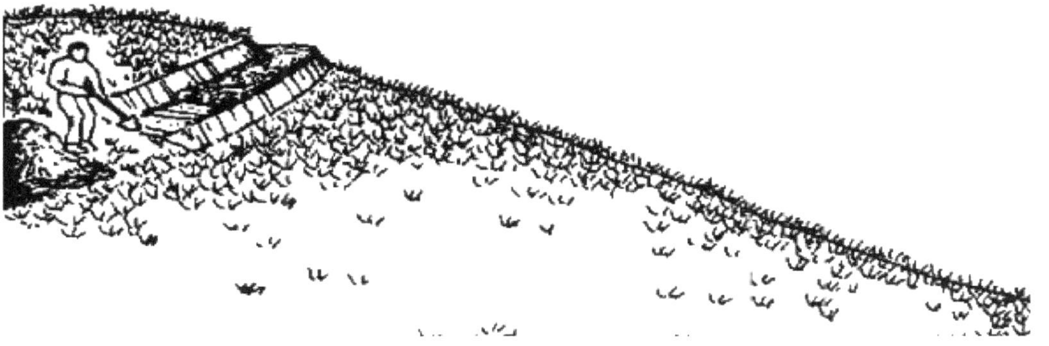

5. Work progresses downslope. Work begins at the drainage side of each terrace and progresses sideways.

6. Grass is planted on terrace risers.

Waterways from draining excess water for fields

In order to avoid problems of erosion at the site of emptying and to reduce the speed of watershed runoff following rains; soil conservation structures should be designed for water retention and infiltration whenever possible. If it is judged necessary to drain water from a field, special care should be taken in selecting areas in which to deposit all the diverted drainage water. Possible drainage areas include pasture areas with a thick ground cover, orchards, or forested areas; where infiltration of the diverted water can probably occur with a minimum of erosion, especially if the water is spread over a large area. Existing waterways may also be used as drainage sites, although one should avoid exaggerating erosion problems by diverting water into areas of active gully formation. In all of these cases, if erosion problems are noted upon diverting runoff water, then a permanent, protected site for receiving runoff should be designed and constructed as soon as possible. Protected drainageways can be formed by reshaping natural drainageways or digging artificial drainageways of a low, broad shape, protecting them from erosion by lining with rocks, planting grass, and/or placing drop structures or check dams periodically.

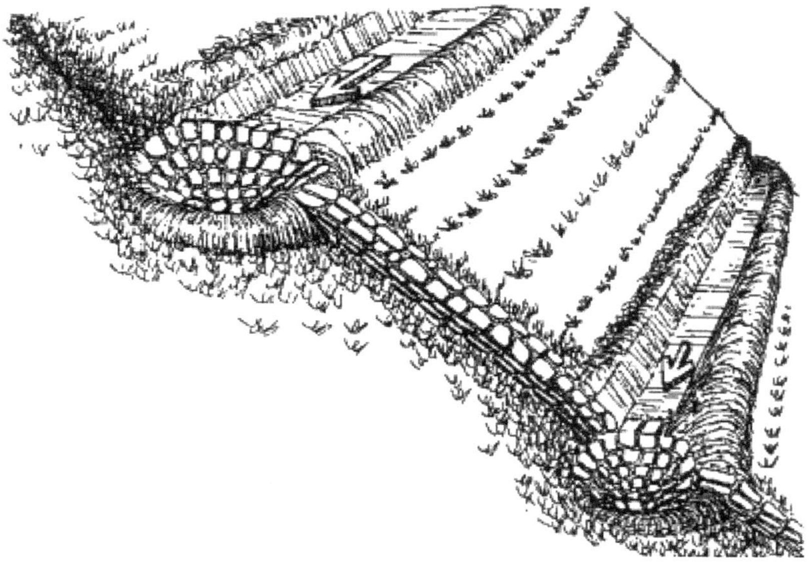

Water drainageway protected against erosion by rock lining.

Waterway design specifications are given in more complete soil conservation guides (Hudson, 1981; Suarez Castro, 1980), allowing the construction of waterways of an adequate size and the selection of an adequate lining method based on local metereological conditions, soil properties, the area of land involved land slope, and the type of protecting lining to be used. If sufficient data is available for the work area, such information can be useful in designing a drainageway of sufficient capacity, without overdesigning it and involving excess work. If however, insufficient local data is available, or it is desired to teach farmers the conservation techniques without discouraging them with complicated tables or formulas, then the appropriate size can be estimated. Hudson (1981) mentions an extremely simplified method ignoring local climatic or edaphic factors: construction of a shallow (30 cm) drainageway measuring one meter wide for every hectare of land area in the drainage basin. While this is probably oversimplified, it does allow one to design drainages of probably adequate dimensions without having to deal with complicated methods. When no appropriate drainage area is available around a field, then retention wells can be dug and water diverted into, and stored in them while it gradually enters the soil or evaporates.

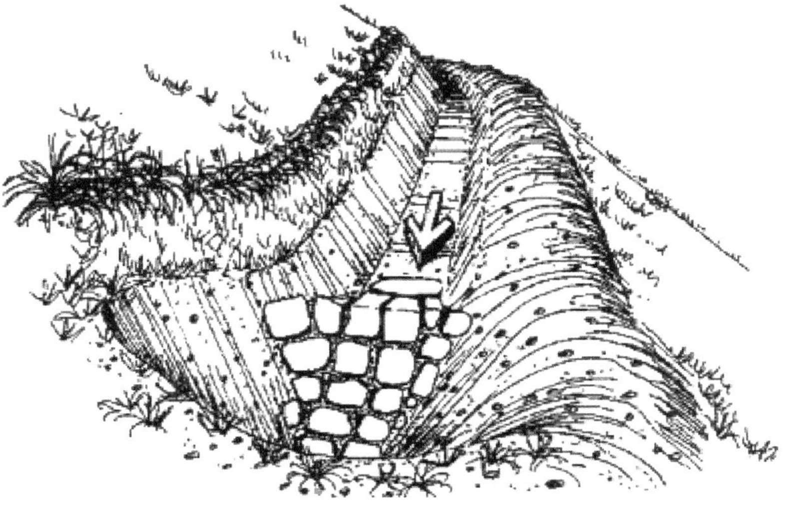

Retention well as site for diverting runoff water.

Special care should be taken to have the waterways and retention wells completely constructed and well vegetated, if grass lined, before diverting drainage water into them. This can be done by preparing them a season or two ahead of time or by preparing a temporary drainage site until the permanent one is ready

Gully Prevention and Control

Gullies form whenever enough water flows in a concentrated area to remove the protecting vegetation and stars digging in and carrying topsoil and eventually subsoil. Some gullies form naturally, others are the result of man's activity, such as redirecting water drainage patterns to protect roads, buildings, or fields. If not controlled, gullies tend to enlarge bath in length and width. Both the soil destabilization and the large amounts of water running through them can eventually ruin fields, roads, or buildings located nearby. When using contour ditching or terracing techniques described in this guide, or carrying out other construction activities requiring the altering of drainage patterns, gully formation can result if the drainageway is not properly protected with rock or grass lining, with retention wells, or with a well vegetated area capable of absorbing the water safely (see previous Chapter an "waterways for draining excess water from fields"). Whatever the cause of gully formation, control measures should be undertaken as soon as possible to prevent further expansion, more damages, and an even more costly control due to the larger size in the future. The most effective control technique is to capture and redirect the source of water responsible. This can be done by means of a gently sloped (1%) contour ditch. This is appropriate only when a suitable adjacent area is available for drainage or if retention wells can be dug to receive the water. Sometimes large livestock watering tanks are filled with water diverted from gullies.

Contour diversion ditch and check dams for use in gully control

Once water no longer enters the gully from above, the soil is often stabilized enough so that revegetation and filling in occur naturally without any further care. The gully, should however be protected from further damage due to cattle trampling and overgrazing, or cultivation. When it is not appropriate to divert the incoming water by use of a contour ditch; if the gully soils, steep surfaces or climate do not permit natural recovery; or if a mere rapid revegetation and refilling is desired, then check dams can be placed across the gully at frequent intervals. These can be made of rocks, wood, wire mesh, etc., and serve to reduce the velocity of any water still flowing and to trap soil particles helping to stabilize the gully so that revegetation can occur. The gully can also be more rapidly revegetated by planting trees, shrubs, or grasses, rather than waiting for natural colonization to take place.

Steps to follow in designing a conservation plan

As in the implementation of any technology, the best results are usually obtained when each situation and pertinent details are carefully examined. The following list of steps has been found useful in both designing conservation plans and in teaching students about soil conservation.

- Select lot, subdivide if necessary
- Study lot, find out problems with respect to erosion, drainage, drought.
- Select suitable drainage site if necessary

- Decide on position of "linea madre"
- Calculate slope
- Select conservation strategy
- Decide on appropriate distance between structures
- Place marker stakes along the "linea madre" at the appropriate intervals
- Survey and mark contour lines
- Realign stakes along contour lines if necessary
- Carry out conservation practice
- Plant protective barriers if necessary
- Plant desired crop(s) along the contour
- Periodically review structures and carry out any needed maintenance practices.

a. Select lot, subdivide if necessary.

This step includes choosing the parcel of land to work on, noting if it should be subdivided into separate management units, each of which should be conserved separately following the rest of the steps. Subdividing the lot into separate units, each of which a relatively uniform slope has, allows the design of a conservation plan far each of the simpler units separately, rather than the difficult task of designing a single plan for a complicated, convoluted area. It should be noted here that most soil conservation techniques are only practical if maintained over a long period of time. Therefore the lot selected should be owned by or permanently available to the involved farmer.

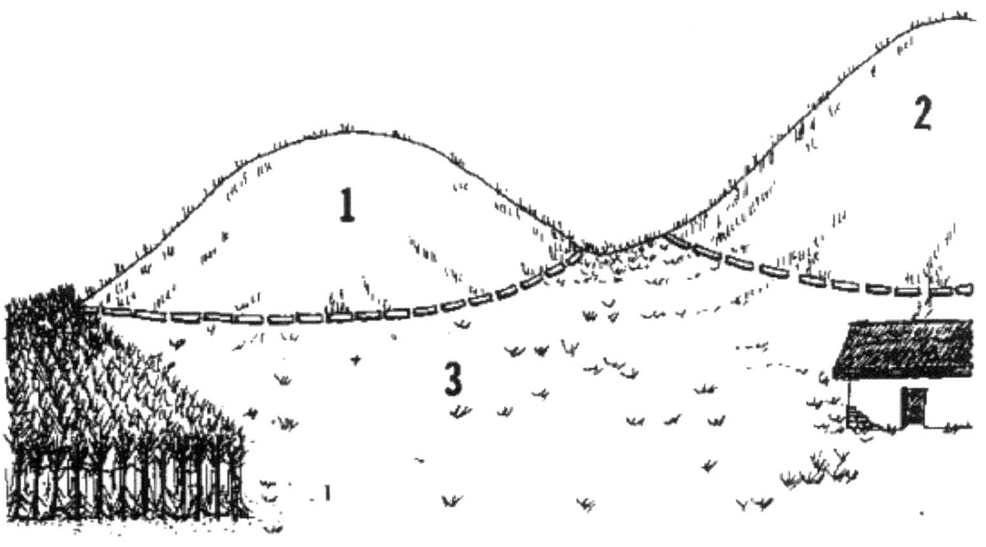

Select and subdivide lot

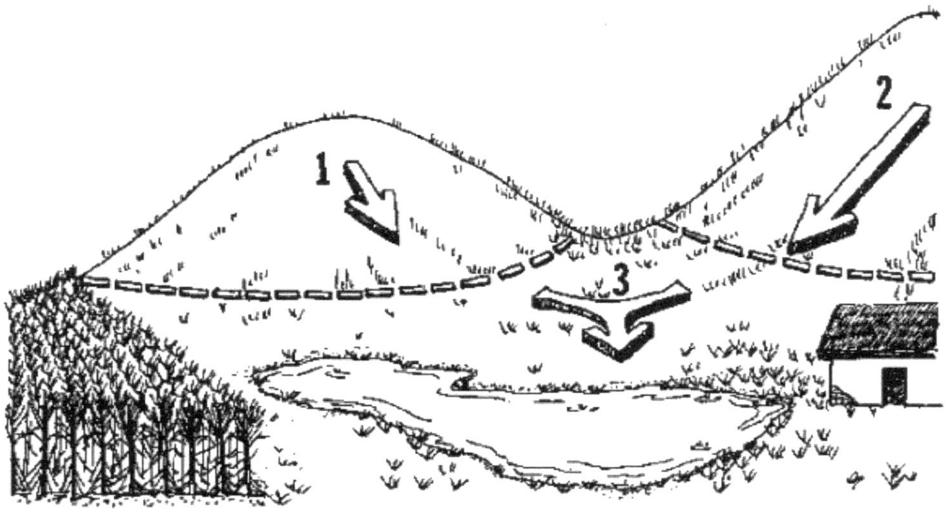

Investigate erosion, drainage, and/or drought problems

b. Study lot, find out problems with respect to erosion, drainage, and drought

This involves asking the farmer about previous years problems; whether flooding or drought are serious problems, the point of entry of runoff water from upper slopes, and sites of serious erosion or drainage problems. It should also include investigating the types of crops, cultivation method used, fallow periods, pests or diseases, etc. Example: In this cornfield, Chapters 1 and 2 are characterized by rapid rainfall runoff, causing small gullies to form. This is followed by excessive soil drying between rains. Chapter on the other hand, is relatively level, but experiences flooding due to poor soil drainage and the runoff.

c. Select suitable drainage site if necessary.

This is an often neglected step which can result in serious problems such as gully formation, flooded roads, or angry neighbors if not considered carefully. If ditches will be constructed to divert excesses of water from a lot, care must be taken to divert the water to a well-vegetated area, a retention well, or to a grass or rock lined waterway. In deciding where to divert excess water, one should also consider the positions of neighboring fields, roads, or houses in order to avoid flooding these areas. If necessary' the construction of drainage ditches or terraces should be delayed until an adequate-site for runoff water can be constructed and thoroughly protected against erosion. (See Chapter on Waterways for draining excess water from fields") Meanwhile, other conservation practices which do not result in concentrations of runoff water, such as the use of contour barriers, maintaining a dense ground cover, etc., can be employed.

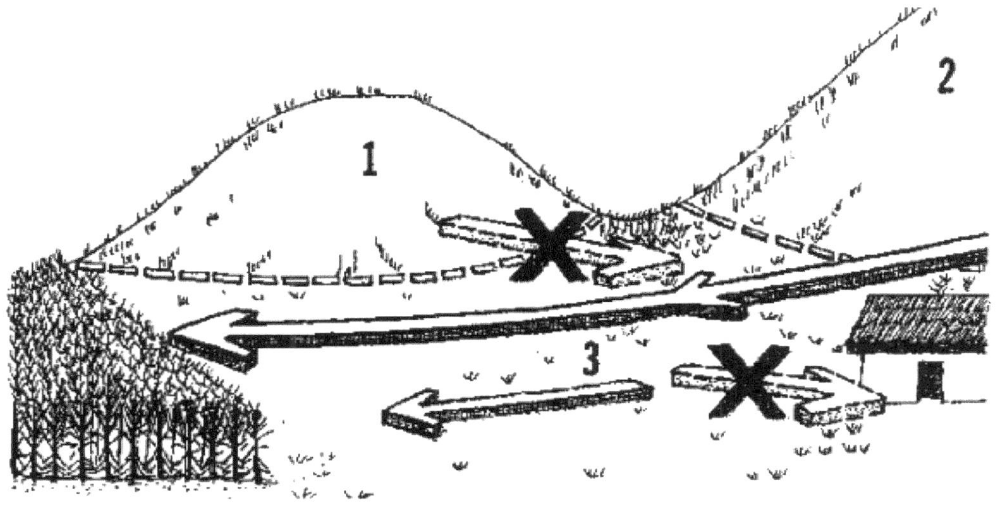

Select suitable drainage site if necessary

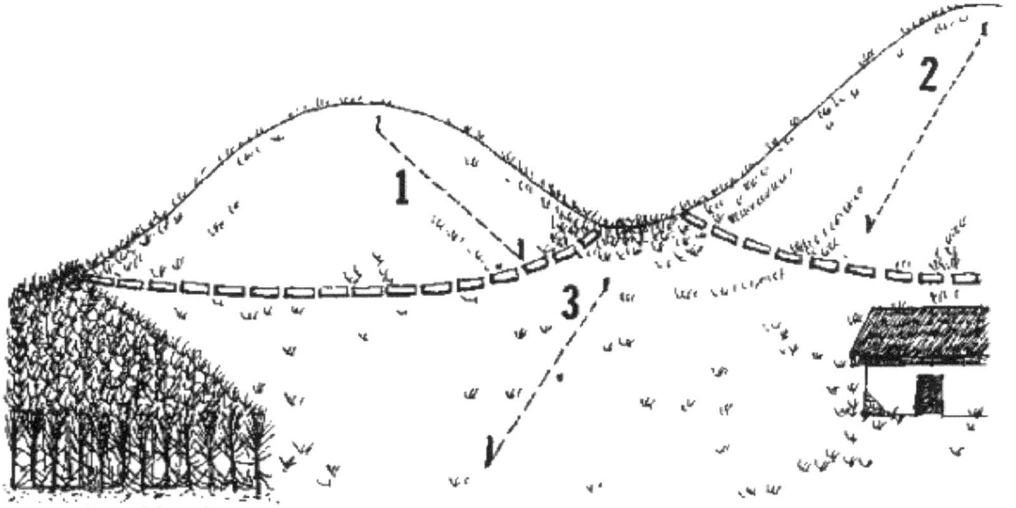

Decide on position of “linea madre”

d. Decide on position of “linea madre”.

This is the imaginary line which best represents the average slope on a unit of land. It will be used as the reference line from which to measure the distance between successive soil conservation structures. It should be positioned so that contour lines will be perpendicular to it.

e. Calculate Slope.

The slope of the terrain should be measured at 5-10 randomly selected sites per hectare and then averaged. The slope can be measured using a string and a small line level as illustrated, or by using an A-frame level the slope of the terrain is important in planning soil conservation strategies. Ideally, flat or gently sloping areas should be used for the production of annual crops, while more steeply sloping areas (35% or more) should be used for perennial crops such as fruit trees or forage, and in very steeply sloping areas (50% or more) the natural vegetation should be maintained. The slope of the terrain is also used when deciding on the distance between successive contour structures, as described in other Chapters.

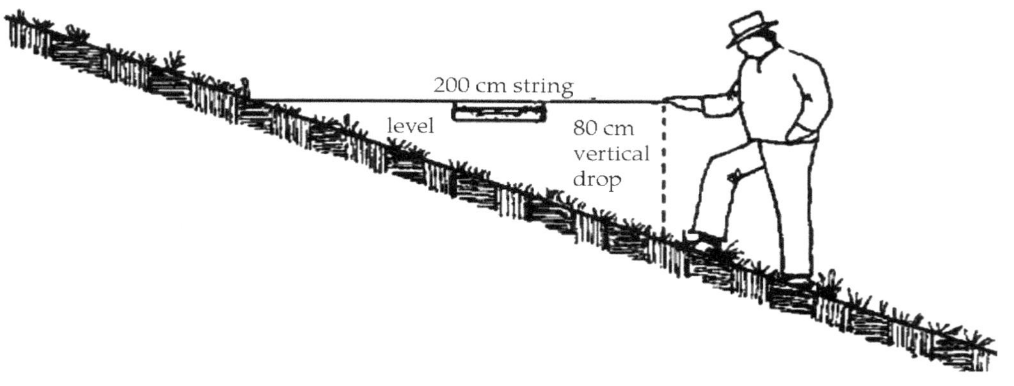

Calculate slope

f. Select conservation strategy

No strict guidelines are presented here for selecting soil conservation techniques due to the tremendous number of variables influencing the decision. However, the following points should be considered as the extensionist and the farmer(s) work out the most appropriate design for an area. 1. Identify resources and constraints which will affect the process. These include the availability of labor, rocks, and live barrier plants; presence of water sources for irrigation; the acceptability of a more long-term investment (perennial crop); the presence of cattle during the fallow period; etc. 2. Select as many complementary soil conservation techniques as possible which are appropriate to the desired cultivation system. EXAMPLE 1: Plant fruit trees on individual terraces protected by contour infiltration ditches and live barriers of king grass. Plant a green manure crop in between the trees and mulch around the tree bases. EXAMPLE 2: Intercrop corn and beans to form a dense ground cover, Construct contour discontinuous narrow terraces with a live barrier of Guatemala grass. Plant a green manure crop such as Velvet bean, during the fallow season. Plant fruit trees on the terraces. 3. Design all structures along carefully measured contour lines (Usually 0%-i% slope). 4. Carefully maintain all contour structures, protective barrier plantings, and drainage ways to endure proper functioning. 5. Review Table, "Advantages and disadvantages of different soil conservation techniques"- and Appendix 2, "Dichotomous key to the selection of soil conservation practices".

Advantages and Disadvantages of Different Soil Conservation Techniques

Technique	Advantages	Disadvantages
1. Protecting Native Vegetation	- avoids erosion and desertification - reserve for native flora and fauna	- often inapplicable due to previous land disturbance - unavailable for crop production
2. Reestablishing Native vegetation	- reestablishes vegetative canopy protecting soil and water resources - restores wildlife habitant - provides useful natural products: wood forage, medicines, etc.	- labor cost to plant and maintain, unless reestablishes naturally - unavailable for traditional crop production
3. Perennial Crop Cultivation Systems (Pasture, Fruit trees, Agroforestry systems)	- avoids annual soil disturbance - once established, low maintenance requirement to produce crop	- unavailable for annual crop production
4. Minimum tillage or mulching in annual crop cultivation systems	- protective ground cover reduces erosion - avoids some pest problems - mulching decreases weed growth - maintenance of soil fertility, structure and organic matter content	- possible pest refuges - possible weed competition - labor intensive - by itself does not control water runoff on sloped land
5. Crop rotation	- maintenance of soil fertility, structure, and organic matter content - breaks pest and disease cycles	- By itself does not control water runoff on sloped land - acceptable alternative crops must be available
6. Contour live barriers	- useful product (forage, fruit, wood) - decrease water velocity and trap soil particles - gradually form terraces behind barriers - little labor and only slight soil disturbance	- appropriate plant must be available - possible shading or root competition - do not stop flow of water - if incomplete germination, ineffective soil traps - may require fencing out of cattle - after planting, ineffective during first rains
7. Contour dead barriers (Rock walls)	- removing rocks makes tillage easier - decrease water velocity and trap soil partioles -if sufficient height, form bench terraces as soil fills in - very little soil disturbance - if planned and constructed during fallow period, function for first rains of growing season	- labor cost - only where sufficient rocks - ineffective if soil fills in level with top - do not stop flow of water - may require fencing out of cattle
8. Contour planting beds and furrows	- series of barriers stop water flow - can be designed to drain or retain water or for furrow irrigation - avoid compaction of planting surface	- labor cost - extensive soil disturbance, susceptible to erosion on slopes unless protected by other structures

9. Contour ditches with protective barriers	- stop flow of water and soil particles - can drain or retain water - simple to design and build - barrier provides useful product	- labor cost to dig and maintain - disturb soil and drainage patterns, may lead to erosion if improperly designed or constructed, especially if draining water - may fill in if barrier not well established - require fencing out of cattle
10. Discontinuous Narrow terraces with protective barriers	- stop flow of water and soil particles - level platform for planting - can drain or retain water - barrier provides useful product - when well constructed require less maintenance than ditches	- labor cost - steeper slopes require deeper "cut", increasing labor - more complicated to build than ditches - disturb soil and drainage patterns, may lead to erosion if improperly designed or constructed, especially if draining water - may fill in if barrier not well established - riser may erode if not well compacted and covered with vegetation - require fencing out of cattle
11. Individual terraces with contour ditches or discontinuous narrow terraces	- level platforms for planting - stop flow of water and soil particles - barrier provides useful product	- labor cost - ditches or terraces must be well maintained to control water movement - may lead to erosion if not well compacted - only applicable for tree crops or agroforestry systems - require fencing out of cattle
12. Continuous bench terraces	- stop flow of water and soil particles - planting of all crops on level platforms - can drain or retain water or carry irrigation water - grass on risers useful as forage	- high construction cost - only practical where year round cultivation possible or where high value crop grown - on steep slopes or shallow soils, terrace width and useful planting space are very small - extensive soil disturbance, highly susceptible to erosion if not properly designed and constructed (must be well compacted with risers protected by vegetation) - require fencing out of cattle

g. Guidelines in determining the distance between soil conservation structures

The most simple rule to follow when placing a series of barriers or ditches in a field is that of locating one structure with every vertical drop of approximately 1.5 m. By standing at the site of one structure and sighting along an extended arm, one can determine the position of the next highest structure.

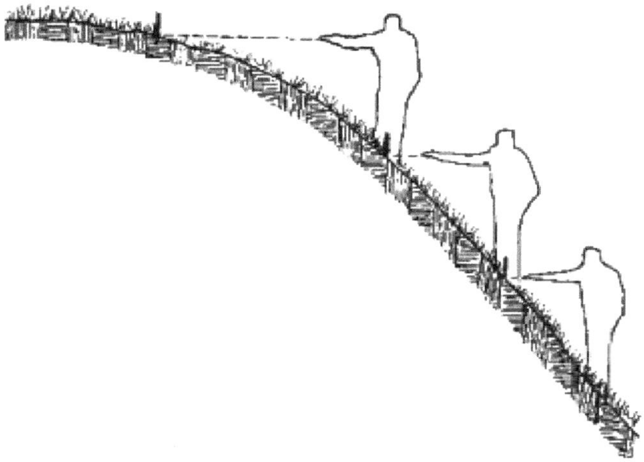

Visual method for determining the distance between successive contour structures.

Technical tables prescribing distances based on the slope of the terrain should be used whenever possible to insure an optimum placement of structures. Simple tables are included in the Chapter on soil Conservation and Water Management Practices" (Tables). Soil texture, soil structure, and rainfall intensity may also affect the optimum placement distance of structures but for simplicity these affects are usually ignored. If the farmer(s) considers the amount of work required to place these structures excessive, then it may be necessary to limit the first seasons work to the placement of every other structure planning to finish the work at a later date.

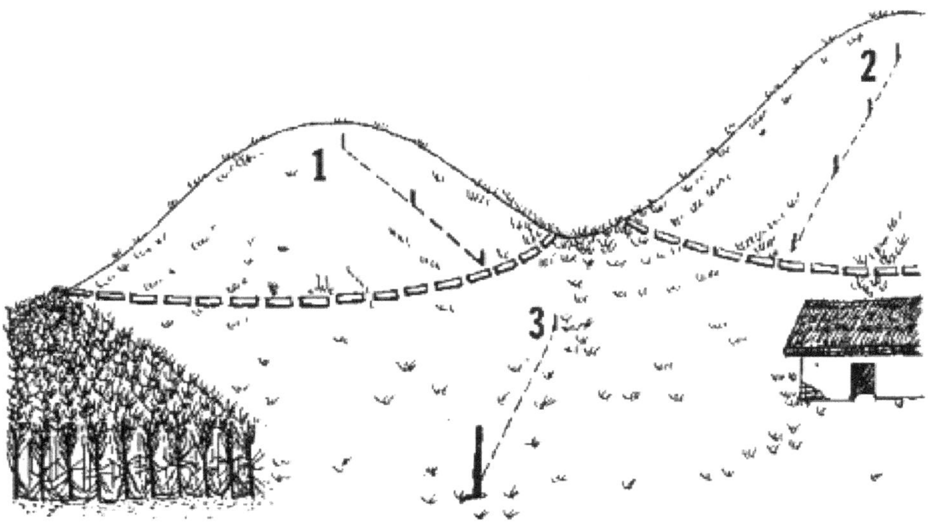

Place marker stakes along the "linea madre" at the appropriate intervals.

- h. Place marker stakes along the "linea madre" at the appropriate intervals. the first stake should be placed along the upper edge of the lot or at the highest convenient place where a structure will be located, subsequent stakes are placed below.

i. Survey and Mark contour lines

This can be done using an A-frame or other type of level. The construction of two types of levels is described in Appendix 4. Care should be taken to carefully calibrate the level and to orient it in the proper direction if surveying lines for drainage ditches.

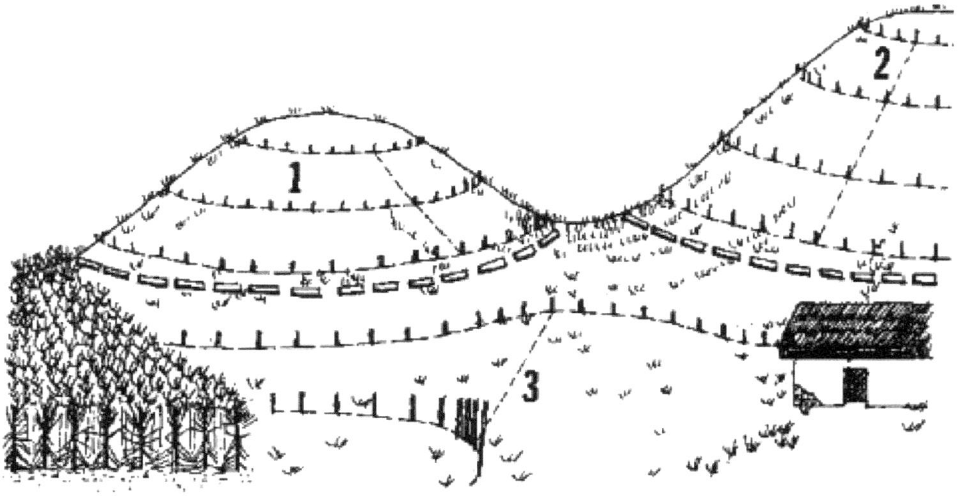

Survey and Mark contour lines

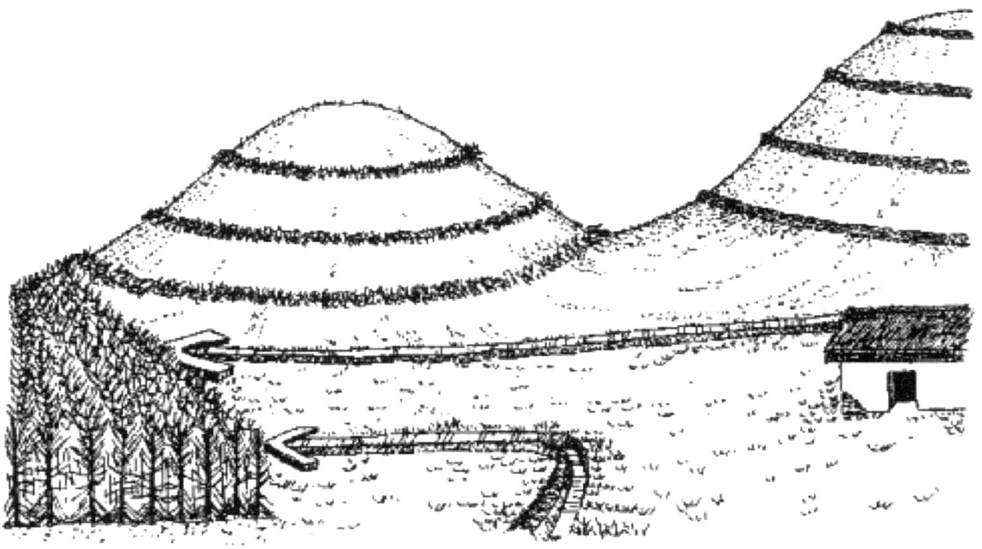

Carry out conservation practices.

j. Realign stakes along contour lines if necessary.

If the contour lines result in an extreme zig-zag pattern which may cause difficulties, for example in plowing a plot, the lines may be straightened somewhat by slightly moving some of the stakes. It should be noted that realigning stakes means that they no longer lie along measured contour lines and that this will have to be corrected for if digging sort of ditches to drain or retain water.

k. Carry out conservation practice(s).

These are described in detail in the Chapter on "Soil Conservation Strategies". l. Plant protective barriers if necessary If not protected by barriers; terraces, ditches, and drainages tend to fill in and cease to fulfill their function as conservation structures. A live barrier planted above these structures acts as a filter to keep soil particles from filling them in and lessens their maintenance requirements.

Plant protective barriers

m. Plant desired crop(s) along the contour

Planting along the contour with closer spacing within rows and more distance between rows forms a partial live barrier with each crop furrow. Care should be taken to plant well selected seeds in order to ensure a more complete germination. On irregular terrain with varying slope, successive contour lines will not be parallel. Point rows are used in the wider intervals in order to fill up the land space and still maintain contour oriented furrows.

Plant desired crop(s) along the contour

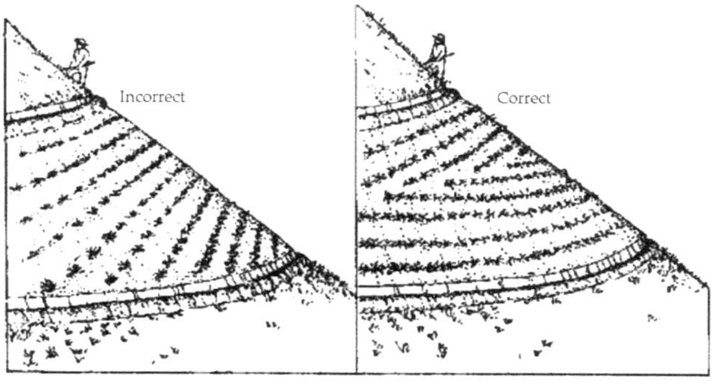

Use of point rows to maintain contour furrow orientation on irregular terrain.

Periodic review and maintenance of structures

n. Periodically review structures and carry out any needed maintenance practices.

Common maintenance practices include replanting of open spaces in live barriers and replanting grass on terrace walls.

SOIL FERTILITY AND ITS MAINTENANCE

Introduction to Soil Fertility

Soil fertility is a concept based on the amounts of essential nutrients available to the plants. The plant nutrients involved are classified as macronutrients such as nitrogen (N.), phosphorous (P), potassium (K), calcium (Ca), magnesium (Mg), and sulfur (S); or as micronutrients such as iron (Fe), manganese (Mn), copper (Cu), zinc (Zn), boron (B), and molybdenum (Mb). The amount of each of these nutrients is not the only factor in determining whether there are nutrient shortages. Soil pH (acidity) is also important because under certain pH conditions, nutrients form insoluble compounds which are unavailable to be taken up by plant roots. A shortage of any of these elements in the soil can slow crop development and reduce yields, something which extentionists should consider when working with farmers to improve crop production. In practice, soil fertility is determined by collecting a soil sample and submitting it for chemical analysis. Since a small sample will be used to infer the characteristics of an entire field, care should be taken to ensure that the sample represents as accurately as possible the conditions present in the plot. Sampling may be done with a soil probe or a shovel, being careful to exclude the surface litter layer and sampling at a depth of approximately 1"-8" below the surface. Subsamples should be taken from random positions throughout the entire field (10-20 per hectare) and mixed thoroughly in a bucket or bag. From this mix, a 1-2 lb. sample is air dried, labelled, and submitted to a laboratory for analysis. A highly variable field should be subdivided into fairly uniform Chapters which should each be treated separately, as they may have very different soil characteristics. Samples may be analyzed with portable field test kits, if available, but it is also advisable to familiarize farmers with the laboratory facilities available to them. In Honduras, for example, soil analyses are performed at the laboratories of Recursos Naturales in Tegucigalpa, Standard Fruit Company in La Ceiba, the Escuela de Agricultura Panamericana in El Zamorano Valley and FHIA in La Lima, Cortes. The most common soil analyses performed are those for nitrogen, phosphorous, potassium and pH, which generally are most important in limiting crop yields. If crops do not appear to respond to changes in these factors then perhaps a deficiency of another nutrient is involved and a mere detailed analysis can be carried out.

Chemical Fertilizers

A simple way to correct some soil nutrient deficiencies is through the application of chemical fertilizers. Their expense may discourage or limit their use in many areas, but applications in small experimental plots are useful in demonstrating their potential and evaluating their feasibility in a specific area. The meet commonly encountered fertilizer are in granular form and contain compounds of nitrogen, phosphorous, potassium, or a mixture of the above. The type of fertilizer and the amount to be used generally are given as recommendations accompanying soil analysis results. In applying such fertilizers, it is important to calculate from the given recommendation the quantity that must be applied in each furrow or at each plant to avoid under- or over-applications. The timing of fertilizers applications is also important. For example in corn, a formula containing both phosphorous and nitrogen is usually recommended at planting time, while nitrogen only, in the form of urea, is usually recommended about one month after planting. Since fertilizer recommendations vary widely based on the soils and the crop to be planted, specifics concerning their use are not discussed here. A mare thorough discussion of fertilizer use is given in "Soils, Crops, and Fertilizer Use: A Guide for Peace Corps Volunteers". Extensionists should familiarize themselves with the fertilization practices in use throughout their zone in order to determine if changes in the types, amounts, and/or timing of fertilizer applications would be beneficial.

Organic Fertilizers

The use of organic fertilizers is another method of correcting soil nutrient deficiencies. Although it is more labor intensive than chemical fertilization, it has the advantages of being locally produced; maintaining soil humidity; improving soil organic matter content, structure and texture; adding a better balance of micronutrients, and improving the soils capacity to hold nutrients. Organic fertilizer refers to a number of types of organic matter which can be incorporated into soils or left as a mulch on the surface. This may be as simple as incorporating crop residues or manure, or as complex as planting a green manure crop, making compost, or raising earthworms to produce materials which will be incorporated into the soil at some later date. A traditional practice in many areas is the burning of residues left when initially clearing land or after a harvest. Although some nutrients are returned to the soil in the ashes, much of the organic matter and nitrogen present are lost to the air. This loss of organic matter affects all of the just mentioned soil properties.

Manures and Crop Residues

The incorporation of crop residues or animal manure is a means of utilizing often wasted by products of the agricultural process. They are relatively simple operations, requiring only the labor to collect and spread manure or crop residues before carrying out the normal tillage operations.

Green Manure Crops

Green manure crops are nitrogen-fixing plants which can be grown for a period of time and then turned into the soil as a source of nitrogen and organic matter. If cultivated for seed production and only turned into soil after dried, they contain little nitrogen and provide mostly organic matter. Table lists a variety of possible green manure crops, all different types of legumes. In crowded areas, where there is pressure to produce as much as possible on every cleared land space, it is often difficult to convince farmers to take land out of food production in order to plant a green manure crop. Fortunately, there are alternatives which allow for the production of green manures without affecting the normal crop production cycle. Some species, such as Gandul, Rice Bean, and others (especially if planted sparsely) can be grown in intercropping systems with other crops. In these cases, the green manure crop can be useful as a ground cover or living mulch which reduces rainfall impact and thereby soil erosion; in reducing weed growth; and in providing useful organic matter, seeds, or forage.

Green manure used as intercropped living mulch.

Some species, such as Kudzu and Velvet Bean are such aggressively climbing, densely foliaged plants, that they are best grown alone. These can be planted during fallow years and then incorporated into the soil Velvet Bean and other drought tolerant specie can even be planted near the end of the growing seas allowed to grow during the dry season, finally to be incorporated into the soil prior to the next planting season

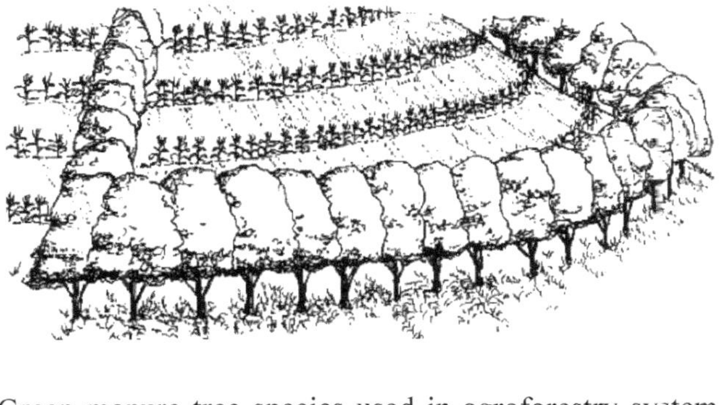

Green manure tree species used in agroforestry system.

Green manure used as fallow season ground cover.

Tree species, such as leucaena, madre de cacao and gualiqueme, can be planted along the edges or in contour strips in the middle of fields, providing a source of leaf material for green manure or forage, and also firewood.

Some Suggested Green Manure Crops

Common Name (s)	Scientific Name	Other Uses	Cultivation	Environment	Comments
Bean, Frijol	Phaseolus vulgaris	Food-pods, seeds	Traditional methods	Many varieties exist for almost all areas	Already commonly cultivated, but often residues not incorporated
Butterfly Pea, Frijol Mariposa	Clitoria ternata	Food-young pods,	2-3 seeds spaced 20-25 cm x 50-100 cm	Warm, sunny, well-drained areas, 380-4200 mm rain, wide ph tolerance 0-1600 m elev.	Perennial, often poor germination, some disease problem in humid areas
Canavalia Jackbean Canavalia Roja, Swordbean	Canavalia ensiformis	Food-pods, seeds, forage	1 Seed 50 cm x 50-100 cm (50-70 kg/ha)	Warm, temperate to hot tropical rainforest areas, sun or shade, 700-4200 mm rain, 0-1800 elev.	Climbing Annual; usually produces seed in 3-4 months; toxic before cooking; mature seeds must be boiled, changing water and peeling seed coat
Clover	Trifolium spp.	Forage	From seed	Developed in temperature climates, possible value in tropical areas	
Cowpea	Vigna unguiculata	Food-seeds forage	For seed: 2 seeds 10-15 cm x 40 cm (25-45 kh/ha) For forage: 2 seeds 8-10 cm x 30-40 cm	Warm climate, 0-900m elev.; Moderately drought tolerant, various soils	Annual, fairly erect, susceptible to insect damage
Crown vetch	Coronilla varia	Forage	From Seed	developed in temperature climates promising for cool highland tropics	
Gandul, Pigeon Pea	Cajanus cajan	Food-seeds Forage Shade living fences	For seed: 3 seeds 30 cm x 30-40 cm (15-45 kg/ha) For forage: 1 seed 5 cm x 30-40 cm	Warm climates, drought tolerant, various soils	Woody, erect perennial (5yrs.), difficult to chop when full grown, 1.5-4 m tall
Gualiqueme, Pita	Erythrina spp.	Ornamental forage, firewood	From seed		Fast growing tree, resprouts from base when cut

Kudzu	Pueraria spp.	Forage	From seed		Very aggressive climbing vine, problem weed in some areas
Lablab Bean, Dolichos	Lablab purpureus	Food-pod, seeds, leaves, sprouts	2 seeds 10-15cm x 50-60 cm (50-70 kg/ha)	Many varieties exist for almost all climates all elevations	Perennial (2-3 years)
Leucaena	Leucaena leucocephala	Forage, firewood	From seed	Hot areas, drought tolerant, lower elev. (up to 500-1000 m)	Fast-growing tree, resprouts from base when cut
Madre de Cacao Madriago	Gliricidia sepium	Forage	From seed or stakes	Hot areas, 0-1600 m elev. 1500-2300 mm rain	Fast-growing tree resprouts from base when cut
Rice Bean, Frijol Arroz	Vigna umbellata	Food-seeds, pods, leaves, seedlings forage	3 seeds, 20 cm x 50 cm. (60-80 kg/ha)	Hot, humid areas, 0-1500 m elev.	Mature seeds in as little as 60 days
Tepary Bean	Phaseolus acutifolius	Food-seeds Forage	1-2 seeds 8 cm x 50-90 cm Seed production: (11-17 kg/ha) Hay producción: (60 kg/ha)	Arid and semiarid regions	Grows poorly in humid areas, salt sensitive, susceptible to insect damage but disease-resistant, annual, semi-erect, 50 cm tall
Velvet Bean, Frijol terciopelo	Mucuna pruriens	Food-pods, leaves, seeds when toasted are like coffee Forage	2 seeds 40 cm x 50-100 cm (10-45 kg/ha)	Drought tolerant, 0-2000m elev. 380-3100 mm rain, soil variable	Trailing or climbing vine, produces great amounts of foliage, can smother weeds, variable life span (up to 6-8 moths to produce seed, use staking when growing for seed)

Composting

Composting is a very labor intensive technique which involves mixing plant residues, manure, soil, and water.

Decomposition takes place over a period of 1-4 months, reducing the mix to rich humus to be incorporated into the soil. Because of the bulkiness of the material and the labor involved, it is most practical in small, intensively worked vegetable plots, but it can also be used as a fertilizer in large fields of any crop if labor is available. A compost pile can be composed of almost any organic materials: weeds, banana stalks, sawdust, coffee pulp, corn or bean residues, etc. Manure from cows, horses, chickens or rabbits, is usually used as a nitrogen source. Ashes, eggshells, and sometimes chemical fertilizers may also be added. Composting can be done with a careful eye on the chemical composition of the ingredients in order to provide an optimum ratio of the elements necessary for the bacterial and fungal decomposers, or by just throwing a little of everything in. Some farmers feel more comfortable with the casual, less exact approach, while others may be mare intrigued by a more scientific attitude towards the process. Some of the types of compost piles are the square aboveground compost pile, the underground compost pile, the volcano compost pile, and the contour compost ditch.

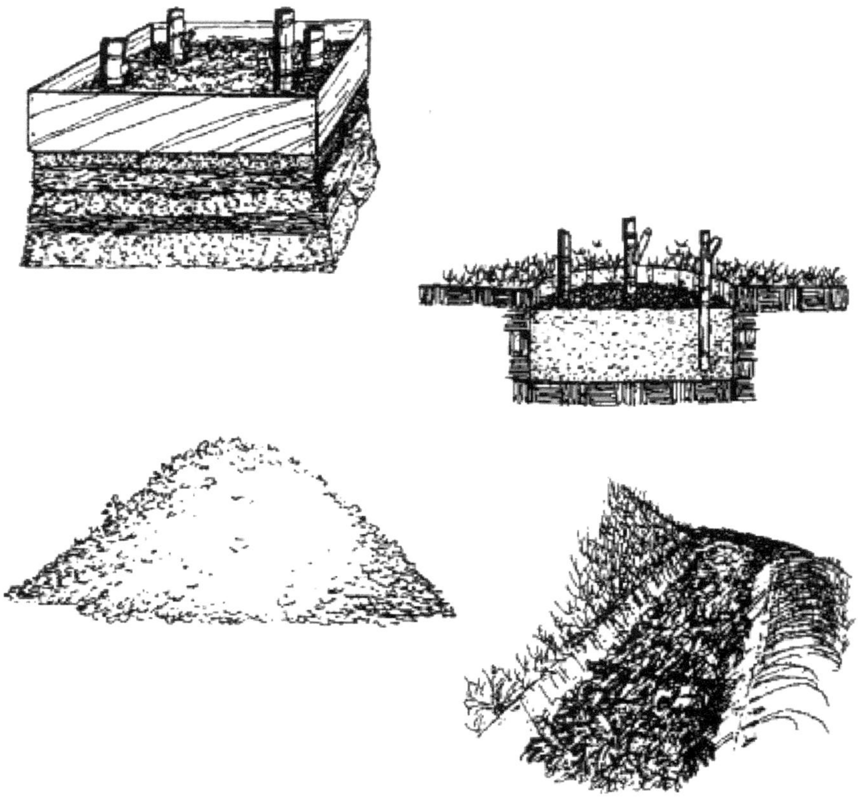

Various Types of Compost Piles

The square aboveground compost pile is tightly compacted into a wooden frame and is often composed of three 12" layers. Each of the layers is made up of 9" of plant materials (both fresh and dry), 2" of manure, and 1" of black soil. Each layer is thoroughly compacted and saturated with water during construction. Posts are placed vertically in the pile (one per m) to be removed after construction to serve as air vents. These piles are generally turned once a month and are ready to use in 3-4 months. The underground compost pile is similar, but is constructed inside a hole dug in the ground (50-100 cm deep). This can be very effective in arid zones or during the dry season, as it reduces the amount of evaporation from the compost. Another type of compost pile is the volcano compost pile, which is simply a mound composed of the same materials (plant matter, manure, soil and water). This type of compost pile can be used to decompose materials much more rapidly (1-2 months), if it is turned and mixed frequently (every 2-3 days).

The contour ditch compost method simply involves filling contour ditches with crop residues, manure, and soil. This method can be very effective on a large scale, leaving the materials undisturbed to gradually decompose over the course of the year, Drainage ditches, should not be filled with compost during rainy periods. Contour retention ditches can be filled however, as the piled up compost serves as a spongy retaining material. The water collected also speeds up the decomposition process.

Composting With Earthworms

Another method of improving soil chemical and physical properties is the raising of earthworms and the subsequent incorporation of both the worms and the rich soil they are raised in. Earthworms can be raised in a mixture of the same materials used in composting, but in this case they are not compacted. They are also often enclosed in a box or tub to prevent their escaping. If a variety such as "Red Wiggler" (Coqueta Roja) is available, it may be much more active and reproduce much faster than other native types of earthworms. The earthworms ingest organic matter, digest it, and leave behind a mineral rich feces which provides nutrients for plant growth. Also by physically passing through the soil, earthworms increase soil aeration and drainage. For a more thorough discussion of composting and of composting with earthworms, see The Encyclopedia of Organic Gardening

EXTENSION METHODOLOGY

The technical aspects of solving problems associated with hillside farmland are much more simple and straightforward than the extension work aspects. Even though there are many variables which affect land use decisions, there are many more which affect the receptivity of farmers in an area to an individual Peace Corps Volunteer's promotional efforts. The possibilities available to a Peace Corps agricultural extensionist are almost endless. Presented here are some sample work activities, guidelines for evaluating work, motivating techniques, and types of groups with which one may be involved. Volunteers should find this material useful in designing and carrying out a more realistic and successful work plan. For additional references on the extension process in rural areas see: Two Ears of Corn, Bunch 1985; Communication Strategies, Lionberger and Gwin, 1982; Communication Innovations, Rogers and Shoemaker, 1971; Agricultural Extension, Gibbons and Schroeder, 1983 and Helping Health Workers Learn, Werner and Bower, 1982.

Timetable of events associated with a "typical" two year peace corps volunteer service.

1st. 3 - 6 months:

1) Settling in, find and fix up living quarters, meet people and make friends
2) Work an learning local customs and language
3) Get acquainted with work zone, communities, climate, topography, crops, technology, problems.
4) Identify at least three communities as target areas.

5) Plant demonstration lot perhaps in the form of a vegetable garden in each of the communities; in own yard, with interested collaborator, or as a school project. Use live barriers, contour ditches, contour planting beds, organic compost, etc.

6) Become familiar with host agency, its resources, and its policies, meet fellow extentionists.

7) Develop reputation as interested, responsible extensionist by attending host agency and community meetings, visiting farmers and keeping informed of seasonal agricultural activities. Help cut farmers in areas other than soil conservation (pest control, fertilizer applications, cooperative management, etc.).

8) Formation of ideas about best potential work strategies.

9) Look for interested collaborators to try out soil conservation techniques.

2nd 3- 6 months:

1) Maintain reputation as interested, responsible extensionist.

2) Discuss specific possibilities for soil observation work with fax. and other extensionists, begin promotional efforts such as introductory lectures, simple demonstrations (See Apendix 5), field trips, etc.

3) Select at least one collaborator from each of the communities and help them carry out some conservation practices on their own land.

4) Encourage these initial collaborators to help seek out and teach other interested farmers.

3rd. 3- 6 months:

1) Keep demonstration lots well maintained, use them to motivate and to teach others to start more demonstration lots.

2) Keep records to measure success of demonstration lots, publicize successes in the communities involved and at host agency meetings.

3) Make sure host agency extentionists, collaborators and other interested farmers understand techniques used; plan more formal courses, demonstrations, or field trips.

4) Constantly be on the lookout for new collaborators, have the initial collaborators help organize interested farmers into groups to receive visits and demonstrations at regularly scheduled times. Encourage initial collaborators to work as volunteer extentionists with other interested farmers.

4th 3- 6 months:

1) Work with larger number of collaborators

2) Organize farmers so that future extentionists will already have distinct groups to visit.

3) Make sure farmers and other host agency extensionist are well-trained to be able to continue designing and building soil conservation structures.

4) Publicize work based on records kept; publicize at agency and community meetings: local, regional or national fairs; radio, newspapers.

5) Train incoming Peace Crops Volunteers.

Depending on the area, the receptivity of the people, the number of growing seasons per year, etc.' this may already be the end of the two year period. If work has proceeded rapidly, all this may have taken as little as one year. In case any time remains, it can be spent trying additional soil conservation techniques (remember that different techniques reinforce each other when combined); helping develop the farmers into more stable, better organized groups; giving more formal training courses for both farmers and extentionists; expanding the work to other communities; or working on another aspect of agricultural development, such as youth organizations, improving marketing, crop diversification, etc.

Guidelines for Evaluating Extension Work

1) Guidelines for evaluating soil conservation techniques used: If records are kept, then a comparison of budgets and crop yields can be made (See Appendix 3). This type of record can be very important in convincing farmers or other extensionists to try similar techniques in another area. Records can also be used to evaluate which techniques are most appropriate or which are inappropriate to the area. Remember, however, many of the benefits of using soil conservation techniques are long-term and may not show a dramatic improvement when using short-term observations to evaluate the techniques. 2) Guidelines for evaluating the extension methodology used: Again records should be kept following the growth of a program. This is often measured in terms of number of people or landarea involved in the program. Extensionists should realize that a large program is not necessarily more desirable than a slow starting, steadily growing one. Finding farmers genuinely interested in carrying out soil conservation practices is difficult (especially the first one). It is reasonable to expect slow acceptance of the techniques at first. Each farmer adopting soil conservation techniques may lead to one or two additional farmers adopting the techniques during the next planting season. Assuming two additional highly motivated new collaborators for every current collaborator, the following growth sequence can be generated:

1	(+2) = 3	(+6) = 9	(+18)=27	(+54)+81	(+162)=243
First planting season	Second planting season	Third planting season	Fourth planting season	Fifth planting season	Sixth planting season

This growth sequence may represent anywhere from two to six years, depending on the climate and the crops involved. If there is only one planting season per year (an arid zone without irrigation or crops such as coffee, fruit trees), after two years the volunteer is at the point with 3 collaborators. It is important that these 3 people be well trained and highly motivated and that other local extensionists are convinced of the value of the work so that it can keep expanding in the area. In this case, since such a small number of collaborators are involved, the volunteer should attempt to visit several communities so that each one can initiate this growth sequence of adopting soil conservation techniques at the same time. For example: if this can be carried out in four communities then 12 collaborators rather than only 3, are involved. If there are two planting seasons per year (moderate climate where a system such as successive corn and bean plantings are practiced), after two years the volunteer may be at the point where 9 or 27 collaborators are involved. If working in 3 separate communities, 27-81 people could be involved. Any of these is enough so that time will be saved teaching in groups. Then teaching groups can be the basis for organizing interested farmers into agricultural committees or cooperatives at a future time. These groups will also form a nucleus of people for future extensionists to visit, reducing much of the time spent in initially getting introduced to a community and discovering interested farmers. This larger involvement also merits bringing in outsiders for field trips to see the variety of work being done and its acceptability to the community. This helps motivate the innovative farmers, making them feel they are doing something important and worthy of respect, and also motivates the visitors to try the techniques in other areas. At this time, the program can also serve as a model for other extensionist to follow. If there are three or four planting seasons per year (year round cultivation possible because of climate or availability of irrigation water), after two years the volunteer may be at the point with 27, 81 or 243 collaborators, depending on the speed in getting started Enough collaborators are now involved that the formation of groups is essential if attention is to be given to all, especially if working with more than one community. These large groups can be powerful forces, making it much easier to attract government aid in the form of courses or loans, and in meriting regional or national publicity for soil conservation techniques me area can now serve as a good model for other agricultural extensionists or development agencies. The group oriented nature may involve changing the work strategy, focusing more on training people in management and organization to stabilize and strengthen the groups, rather than strictly focusing on training in agricultural techniques From this growth sequence, one can see that even though growth starts very slows if it can be maintained at a steady rate, it is soon growing so fast that one extensionist can hardly keep up with it Essential to the process however, is quality and success, especially in the initial period, in order for the techniques to appear acceptable and attractive to farmers

Extension Techniques

A soil conservation extension program almost always faces a challenge initially to generate awareness that soil erosion is a major problem that merits dealing with often farmers do not perceive deforestation or environmental degradation as problems, and may attribute poor production to other factors (lack of credit to buy fertilizers, quality of seed available, lack of modern farm machinery, etc). One of the first goals of any extensionist, therefore, should be to change same of the attitudes farmers may have and introduce some of the following ideas: The importance in an increasingly crowded society of permanently cultivating the same plots of land while maintaining others in an undisturbed state to protect water sources and wildlife, the simplicity of adopting soil conservation techniques, and the advantages of labor intensive improvements as compared to capital intensive (buying fertilizers, improved seeds, farm machinery) improvements. Only upon generating genuine interest in soil conservation techniques can an extensionist expect farmers to be willing to try out and to care for, any new types of structures or planting systems introduced. Several methods are available to extensionist attempting to change farmers' attitudes and generate interest in new techniques. Some of the common ones are informal discussions, lectures, films or filmstrips, classroom demonstrations, demonstration lots, field trips, visits from farmers already using the techniques, financial incentives, and soil conservation courses. Informal discussions with farmers can be one of the extensionist's most effective techniques, especially in more isolated communities and where people may be unaccustomed to receiving courses, attending meetings, or receiving visits from extensionist Informal discussions provide an opportunity to make friends, to have people understand what to expect of the extension program, and to discuss ideas in an informal setting. Friends made in this manner, often turn out to be the first interested collaborators in an area. More formal presentations such as lectures, films and films strips, and classroom demonstrations (see Appendix 5) made to a group of people, allow for the presentation of more information to more people and the use of visual aids to make some of the ideas clearer. They have the disadvantage, however, of requiring people to attend a meeting at a fixed time, something people may be unaccustomed to or very reluctant to do. These types of presentations are probably most effective when several individuals have already expressed an interest during informal discussions. In communities where people are not accustomed to receiving focal presentations, the format and content of the lecture, film, or demonstration should be designed carefully to ensure that the people attending understand how this will relate to their own farm work. Demonstration lots are small plantings, carefully prepared and cared for, that demonstrate some or all of the techniques which are to be promoted in an area. "Seeing is believing", and that is the main advantage of this technique. People are given a concrete example, so that when an extensionist talks about digging ditches in the middle of a corn field, planting in contour curves, or using organic fertilizer, farmers will have a clearer concept of what these terms are describing. They also provide a local trial to evaluate the appropriateness of the techniques under local conditions. Field trips to, or visits from farmers already using soil conservation techniques provide an opportunity to evaluate what is being done in other areas and to consider their appropriateness in the new area. These are especially valuable if there is a chance to discuss the new techniques directly with the farmers involved. This will permit a more thorough consideration of tine and labor involved and the rewards to be expected. Agricultural extensionists should encourage their collaborators to seek out and share their experiences with other farmers in the area. Extensionists might even consider making a "moral contract" with collaborators, requiring them to teach two additional farmers, who in turn will each promise to teach two more farmers. etc. In this way the number of farmers learning and using the techniques increases more and more rapidly with time. The extensionist should be aware that if the model area or farmer has received any special attention or aid to carry out the work, this may be interpreted as a prerequisite to the success of the technique. Financial incentives (credit, seed, fertilizer, food for work, etc.) are available from many national and international agencies interested in rural development projects. These often are available only for groups of farmers. They can be used to attract participants to a project designed, for example, to bring people together for formal classroom sessions where new techniques are described, followed by carrying out the practices on their own land. In this process, the involved farmers receive benefits from increased production and a longer useful life of their fields. In Appendix 3, an extension program is described in which fifty farmers were involved in such a project. If carried out throughout a region or country, these types of projects can benefit the economy because of greater self-sufficiency in production and can reduce migration to urban areas and avoid often-destabilizing political pressures for rapid land reform measures. This method has the advantages of attracting a larger number of people to be trained during the extension program and the more immediate achievement of economic benefits to a larger number of people. A possible disadvantage of this strategy is that it may overlook the importance of future acceptance of soil conservation techniques by farmers not involved initially. If the same motivating benefits are not available to other farmers, they may not feel that the soil conservation work by itself is worthwhile. Depending on the receptivity and the subsequent extension methodology followed, however, this may not present a problem. In fact, the more immediate high visibility of such a program may allow a great increase in the effectiveness of a Peace Corps Volunteer during a two year service. The high visibility of such a large project could also be important in providing a site for field trips to motivate or train other farmers and extensionists, and a basis for publicizing the success of soil conservation techniques on a regional, national, or international level. Soil conservation courses allow extensionist to teach a variety of techniques to farmers, more than they may be able to learn by doing their own field work. It should be remembered however, that many people are unaccustomed to learning in a classroom format, and that the courses should involve as much practical fieldwork as possible. Courses are probably most effective in training extension workers, on once several farmers in an area have already tried some soil conservation techniques on their own fields and seem receptive to learning more.

Working With Groups

As an extension program gets under way in a community and starts growing, it becomes more and more necessary for the extensionists to work with groups as time may not always be available for individual visits. If working in communities with no organized groups, extensionists are faced with the task of initiating groups with which to work. It is not unusual for community members to be resistant to the idea of joining such a group. For many reasons (a tradition of working alone, local feuds or jealousies, suspicion of unwanted economic or political commitments, etc.)) this may be one of the extensionists most difficult tasks. dealing with and overcoming this reluctance is an individual matter and every extensionist works things out uniquely. It takes time and communication to develop a relationship that both the community members and the extensionist are comfortable with. Other people, including other agricultural extensionists, nurses, or school teachers, who have had experience working with similar community groups, may have very useful suggestions in these matters. Although presenting new information to farmers is more efficient on a group than on an individual basis, this may not be so for actual field work. The extensionist must evaluate each community to decide if it is more appropriate to carry out demonstrations on individually or group tended lots. In group cultivated lots, the work and risks are shared, lessening the burden of any individual. However, sometimes farmers are less conscientious about caring for a crop if it is not theirs individually. In such cases, it may be most appropriate for the extensionist to present new information or techniques at a central location, such as a school, a small parcel of land made available to the extensionist for this purpose, or by rotating among the individual parcels of the group members. After the group presentation, the extensionist should, whenever possible, discuss or visit each individual's parcel to ensure that the technique will be put into practice correctly. Organized groups of farmers can also be very important in the continued spread of the introduced techniques after the extensionist leaves. They increase communication among farmers, they serve as support groups for innovative farmers, they are more likely to attract the attention of other extension programs, and in some cases they manage credit funds which make it easier for farmers to implement certain technologies. Extensionists should familiarize themselves with the different types of groups so that they have a better understanding of how groups may facilitate, or possibly inhibit, the continuance of their promotional effort. Agriculture committees are often informal assemblages of people drawn to meetings by a common interest in agricultural innovations. These committees are very flexible and allow for the admission of new members or formation of new committees upon demand. This can be a big advantage when working in a new area as the success of the program will probably attract more and mere interested persons. This flexibility also leads to the danger of dissolving rapidly if interest wanes, especially if dominated by only one or two enthusiastic members. The members of these committees can, however, be very important resources as local volunteer extensionist singe very often they are attending solely based on interest and a desire to learn, without any other reward. These committees can also be a good starting point for the organization of an agricultural cooperative if desired. Cooperatives are more formal groups, generally organized around a set of by-laws or constitution, requiring members to fulfill certain responsibilities (pay dues, attend meetings or workdays) and granting them certain privileges (credit, buying and/or selling at favorable prices, right to farm a certain portion of land). Because of their formal nature, cooperatives are likely to be more permanent organizations than agricultural committees. Also as a result of this more formal nature, they are less flexible about admitting new members. Many times cooperatives receive preferential treatment from. governmental agencies when soliciting training courses, credit, or other types of assistance. There are several types of cooperatives, differing in the nature of the rights they grant and the responsibilities they require of the members. Production cooperatives involve the members working together in the production process, such as farming the same piece of land. Credit cooperatives manage a common fund which is used to loan members money, an alternative to more expensive and often unavailable bank loans, Buying and selling coops pool all the farmers buying orders, buying in bulk for cheaper prices, and pool the farmers' produce to sell an higher prices or lessen transportation costs Soil conservation extensionist may also be involved in working with other types of groups, such as schools, youth groups, housewives groups, etc. Regardless of the type of group an agricultural extensionist works with, some basic concepts must be considered. First, groups should be goal-oriented or purposeful, that is they must provide some advantage to their members; some incentive to spend their time with the group Second, the group must have a structure, organization, plan of activities, and disciplinary code which permit the attainment of its goals. Third, care must be taken in planning, promoting, and realizing all group oriented activities to avoid disillusion among members and abandoning of the group. This may result from joining a group without understanding its stated goal or joining a group incapable of attaining its stated goal because of flaws in the design of its structure, organization, plan of activities, or disciplinary code. When these concepts are kept in mind, then the group is much more likely to serve its members in a productive, self-sustaining fashion Once confident of the usefulness and power of their own group, community members will be much more motivated to work within the group framework to improve their own situation.

This guide attempts to provide agricultural extensionist with an understanding of some techniques which have been found helpful in improving hillside farming systems. Hopefully each extensionist can find one or several techniques presented here which are appropriate for any specific work area. Clearly it is not the role of this guide, just as it is not the role of the extensionist, to dictate the techniques to be used in any given situation. Rather agricultural extensionist and farmers should realize the realm of possible techniques available to them, enabling them to formulate their own plans according to the many variables (climate, topography, labor, markets, etc.) which determine the appropriateness of any given farming practice for an area. Furthermore it is hoped that hillside farms are not to be regarded as "things to be fixed once and for all" with a set of soil conservation techniques. Farms, like individuals and communities never reach an ideal state of development, rather they are dynamic units which must adjust over time to a changing climate of environmental, human, economic, and political factors; all of which require that new practices constantly be tested and evaluated. If agricultural extensionists can communicate to farmers the importance of testing and evaluating new techniques, such as the ones included in this guide and many others, then both will be better prepared to address the problems facing the farmers now and in the future. The technologies described in this guide are all simple options available to small-scale hillside farmers in the use of their land. It is hoped that through the use of these practices that better yields may be sustained for more years on each cultivated plot of land. In addition to the increase in crop production, the introduction of soil conservation techniques may facilitate the promotion and realization of more long-term effects of area wide conservation programs; such as the management of watersheds to maintain a reliable, clean, drinking water supply and the protection of undisturbed lands as reservoirs of native flora and fauna.

REFERENCES

1. Agpada, A. et. al. 1975. Manual of Reforestation and Erosion Control for the Philippines. Compiled by H. J. Weidelf. German Agency for Technical Cooperation Ltd. Eschborn, W. Germany.
2. Bunch, Roland. Two Ears of Corn: A Guide to People-Centered Agricultural Improvement. 2nd Edition.World Neighbors, Oklahoma City. 1985.
3. Duke, J. A. Handbook of Legumes of World Economic Importance. Plenum Press, N. Y. 1981.
4. Firewood Crops: Shrub and Tree Species for Energy Production (2 vol.) National Academy of Sciences. Washington, D.C. Vol. 1. 1981 Vol. 2, 1983.
5. Gibbons, Michael J. and Schroeder, Richard. 1983. Agricultural Extension Manual. M-18. Peace Corps Information Collection and Exchange.
6. Greenland, D.J. and R. Lal (eds.) Soil Conservation and Management in the Humid Tropics. John Wiley and Sons, New York. 1977.
7. Hudson, Norman. Soil Conservation. 2nd Edition. Cornell Univ. Press, Ithaca, N. Y. 1981.
8. Kunkle, S. H. and J. L. Thames (eds). 1977. Guidelines for Watershed Management. Conservation Guide #1. FAO, Rome, Leonard, Dave. 1969. Soils, Crops, and

Fertilizer Use: A Guide for Peace Corps Volunteers. Peace Corps Program and Training Journal Reprint Series No. 8.

9. Lionberger, H. F. and P. H. Gain. Communication Strategies: A guide for Agricultural Change Agents. The Interstate Printers and Publishers Inc. Denville, Illinois. 1982.

10. Michaelson, Tage. 1980. Manual de Conservación de Suelos pare Tierras de Ladera. Ordenación Integrada de Cuencas Hidrográficas. Proyecto PNUD/FAO/HOND/77/006. Tegucigalpa, Honduras.

11. Rodale, J. I. (ed.) The Encyclopedia of Organic Gardening. Rodale Books, Inc. Emmaus, Pennsylvania. 1971.

12. Rogers, E. M. and F. F. Shoemaker. Communication of Innovations: A Cross Cultural Approach. 2nd Edition. The Free Press, MacMillan Publishing Co., New York. 1971.

13. Sheng, T.C. and H. R. Stennett. 1975. Forestry Development and Watershed Management in the Upland Regions: Jamaica, Lecture Notes: Watershed Management and Soil Conservation Training Course.United Nations Development Programme, FAO, Kingston.

14. Soil Conservation Service. A Manual on Conservation of Soil and Water. United States Department of Agriculture. Agriculture Handbook No.61, 1954. Reprint R-38. Peace Corps Information Collection and Exchange. Manual M-38. September, 1982.

15. Suarez de Castro, Fernando. Conservación de Suelos. Instituto Interamericano de Ciencias Agrícolas, San Jose, Costa Rica. 1980.

16. Tracy, Frederick and Ricardo Perez. 1986. Conservación de Suelos. Folleto de Campo: Distanciamientos y Dimensiones de Obras Físicas. Proyecto Manejo de Recursos Naturales, SAN/AID #5220168. Tegucigalpa, Honduras.

17. ___. 1986. Manual Practico de Conservación de Suelos. Proyecto Manejo de Recursos Naturales. SAN/AID #522-0168. Tegucigalpa, Honduras.

18. Tropical Legumes, Resources for the Future. National Academy of Sciences. Washington, D.C. 1979.

19. Vickery, Deborah and Vickery, James. 1981. Intensive Vegetable Gardening for Profit and SelfSufficiency. Peace Corps Information Collection and Exchange. Reprint R-25.

20. Werner, David and Bill Bower. Helping Health Workers Learn: A Book of Methods, Aids and Ideas for Instructors at the Village Level. The Hesperian Foundation. Palo Alto, California. 1982.

CHAPTER 7

Rice Farming in Upland Areas

Chapter Contents

- **Introduction**
- **Land Selection and Preparation**
- **Soil and Compost Preparation**
- **Transplanting**
- **Harvesting and Soil Treatment After Harvesting**

Agriculture in the Philippines is rapidly changing, as new farming techniques and varieties come into use. The deterioration of the environment and natural resources such as the deforestation, land degradation misuse of pesticides and chemicals and the loss of genetic resources coupled with the fast growing population in the country gave the DAs' challenge in refocusing strategies in ensuring food security and availability.

INTRODUCTION

Rice is the most important food crop of India covering about one-fourth of the total cropped area and providing food to about half of the Indian population. This is the staple food of the people living in the eastern and the southern parts of the country, particularly in the areas having over 150 cm annual rainfall. Rice is life for thousands of millions of people. In Asia alone, more than 2,000 million people obtain 60 to 70 per cent of their calories from rice and its products. Recognising the importance of this crop, the United Nations General Assembly declared 2004 as the "International Year of Rice" (IYR). The theme of IYR—"Rice is life" reflects the importance of rice as a primary food source, and is drawn from an understanding that rice-based systems are essential for food security, poverty alleviation and improved livelihood.

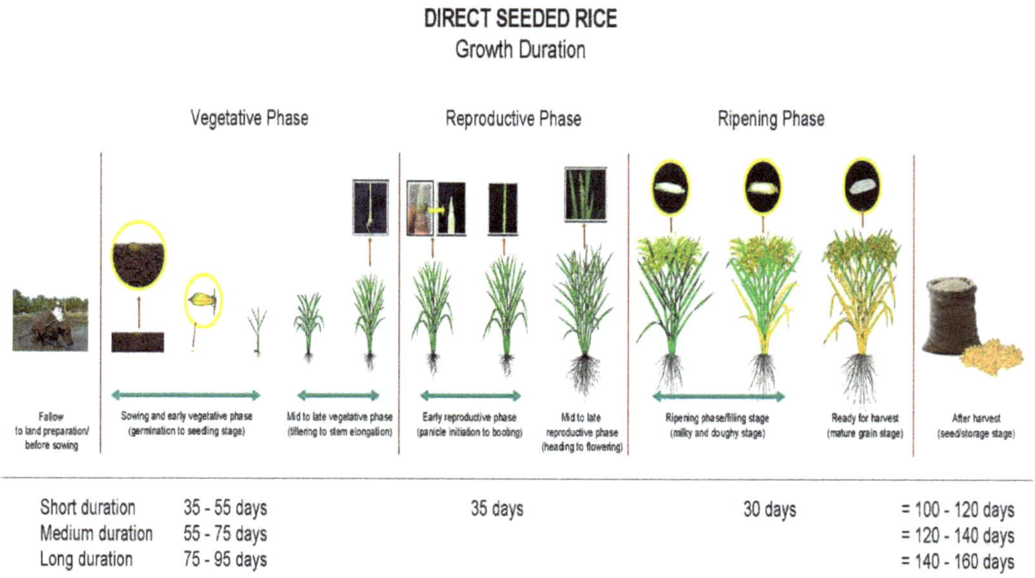

Illustrative Pictorial Stages of a Basic Rice Cultivation Cycle (4-5 months in duration)

A basic rice cultivation cycle includes land preparation, soil and seed-bed preparation, sowing of seeds, transplanting seedlings, general crop growth management practices such as weed control, application of composted organic matter for boosting soil productivity, water control, insect pest monitoring and control, harvesting, threshing, winnowing, drying, storing, selecting and collecting seeds for next cropping, and eventually milling the paddy for family consumption. The key to sustainable rice cultivation is for every farmer or rice planter, at each harvest of rice crop, should actively select and harvest enough rice grains as seeds for the next cropping. When harvesting rice grains for seeds, the area of land for the next crop can be decided at this time and the adequate amount of grains as seeds can be safely set a side. The Handbook on Upland Rice Farming basically follows basic rice cycle as illustrated due to its simplicity; it can be easily integrated into the PNG's traditional root, tuber and banana crop farming system.

LAND SELECTION AND PREPARATION

Land Selection

The first step is to select a land to cultivate rice. Upland rice cultivation is basically rain-fed farming practice. So it is also important to select an area with adequate rain-fall throughout the year. The land with soft-top black soil with more than 2 cm is preferable.

It is preferable to select a site that has flat to middling rolling land ground surface rather than on steep, sloping and mountainous sites. Flat or mildly rolling land will have little or no problem with soil erosion and will hold the soil containing essential plant nutrients when there are high or intense rainfalls. If it is difficult to find a flat land, try to find an area that is not too steep. If you need to use a slopping site such as on mountain sides, you will need do some soil digging or excavation to build terraces where good gardening or organic surface top-soils can be filled onto the excavated material. Well-constructed terraces can hold the soil to hold essential plant nutrients, organic matters and even saturated soils or rain-water that can prevent plant nutrients and soils losses through soil erosion or land-slips. When thinking of the size of farm, you need to consider the available resources such rice seeds and man power. When you decide on the new garden and how big the garden should be, remember to take into consideration your available resources and inputs for rice cultivation. The first consideration is your rice seeds, then the family labour you have, and the tools and the means by which you will cultivate your land. If you are going to cultivate rice for a first time, it is better to start with a area that is 25 m^2 (5m x 5m) or less, and this will require only 30 grams of rice seeds. It is preferable to select a site near or not so far from one's house or dwelling place (village) so that it is easier to manage and give a close attention to your garden. If you have somebody who can help you to manage your rice garden such as your relatives and sibling, you can start with the bigger area. If you have 0.5 kg of rice seeds with a good germination rate (80%), the quantity of rice seeds and germination rate can enable you to plant garden as wide and as long as 20 m x 20 m giving you a total crop area of 400 square metres (400 m^2).

Land Preparation

After you have selected your rice garden site, you can now plan to clear your land at the time of year as indicated in your cropping calendar. Clearing of thick wooded forested area may take longer time to clear so consideration should be given for their felling and clearing. Soft wooded forest may take shorter time to clear and for the grass and broad-leafed site, it may even take much shorter time to clear. All forest tree twigs and leaves, brush, shrubs and grasses from the cleared sites can be utilized for composting and are excellent source of organic matter for composting and soil improvement when they are incorporated into the rice garden soils. All cleared plant materials from the rice garden sites should not be thrown away. Burning of dried plant matters should be discouraged as well as. Burning of dried materials from the forest, brush, shrub and grass clearings will destroy all organic matter and in most cases will destroy the soil structure and its fertility. Sites cleared for rice cultivation should be devoid of all solid materials that would obstruct planting of the rice crop. Where the soil is been tilled with hoeing or ploughing, the soil surface can be manually evened out with hand-tools or leveling equipment.

A farmer clearing the land by slashing grass and wood shrub.

SOIL AND COMPOST PREPARATION

The next step after selecting the garden site and preparing the land is the soil preparation, which is one of the most important steps The condition of soil influences the fertility of land, presence of pests and diseases, and eventual yield. Therefore, the simplest and effective way of obtaining a good condition of the soil is to plough the available organic materials or green manures or composted organic matter into the soil. Weeds, kitchen wastes, home-yard shade tree leaves, including the previous crop stubbles, and rice husks, can be used and ploughed into the soil for adding plant nutrients into the soil.

Soil Preparation

Soil preparation should take place one month before transplanting in order to give enough time (1 month) for organic materials or plant or animal materials to de-compost completely and become fertilizers in the soil. Basic steps up to digging trench, which can be done in a day, are indicated below.

Hoe the soil to a depth of Plough 20 to 30 cm depth to mix soil itself (plough 30- 40 cm depth when using a machine)

Till all grass or rice Plough straw, weeds, any soft plants like banana stem and leaves , and kitchen wastes (without seeds or flowers) into the soil

Garden should look like this after tilling or ploughing with organic materials

Make a trench by digging the surrounding of the garden (10 cm depth) and putting soil back to the garden for covering green manure with soil. Make a gate to release water for water control.

Soil after tillage and use of compost would be friable and easily crumbles in one's hand after a month

Compost Making

Compost should be used as an organic fertilizer and must be applied one month after sowing of the crop. The ideal time to make compost is during the garden site clearing and soil preparation. The reason for doing so is that compost making is similar to soil preparation. It is good to prepare compost right next to the rice garden to make it easy to create and apply compost to the garden. Basic steps for compost making are described below. (1) Secure a site and designate an area for making a compost-heap. (2) Dig the borders of the designated area for the compost-heap digging to about 10 cm depth or place logs to differentiate the compost area from the other areas. (3) Collect rice straws and other organic materials (green manures) such as weeds, any soft plants parts like banana stem and leaves, kitchen wastes, etc. Weeds with seeds and flowers can be also put into compost because the temperature inside the compost will be high and hot enough to destroy most of the seeds and will not germinate. (4) Lay the rice straws on the entire heap area as the first-layer (5) Put a layer of plant materials or organic matter on top of the first-layer of straws. (6) Sprinkle some soil over the plant materials, and build a thin-layer of soil. This soil layer should not be more than 1 cm in thickness (less than 1 cm in thickness). (7) Even out the layer of the soil by hand and by using a straight-edge implement or stick. Slight down-down pressing can be done on the soil layer by foot. (8) Repeat (5) to (7) to make three to five layers of green manures and thin layer of soil as illustrated (9) Sprinkle some water to make 50-60% wet-basis moisture content. An indication of 50%w.b. moisture content in the compost is achieved when you grab and squeeze it in the palm of your hand and be able to make a mass of earth or soil without it crumbling quickly. An indication of 60%w.b. moisture content is when there is visible small amount of water squeezed out of the soil. Further illustration of soils having 50-60% moisture content (w.b.) (10) Cover the whole compost with banana leaves to promote de-composting to take place. (11) Turn-over the compost heap after a month, and add some water if the heap is too dry and cover the heap again with fresh banana leaves again. The internal temperature of compost should be higher than 40°C at the peak stage of compost activity. It is preferable to turn-over the inner materials to that of the outside from time to time when checking the compost, to ensure even decomposition and break-down of organic matter.

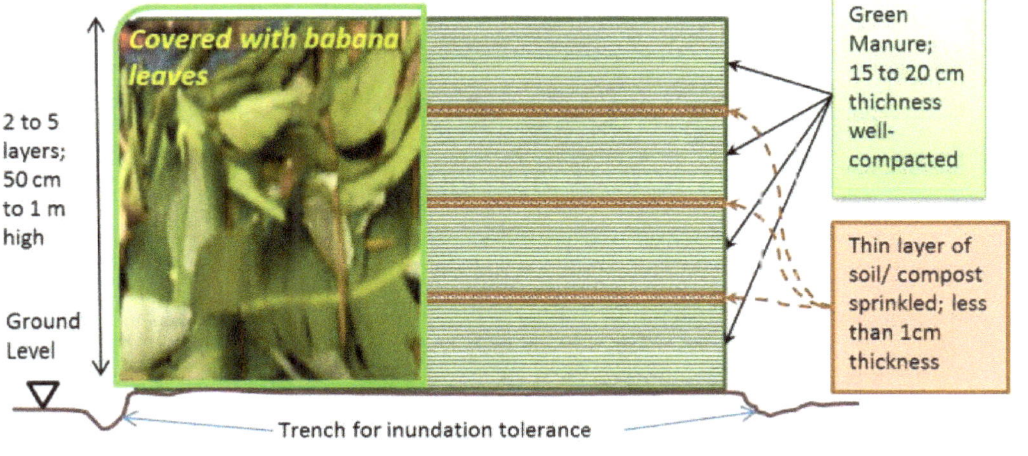

Schematic illustration of a compost-heap construction, showing each layer of material.

Compost-heap being pressed downward with foot as organic matter and soil are added on layer after layer (Step 1-5 in compost making).

Newly completed compost-heap with broad banana leaves covering the top.

A partially decomposed wood and tree twigs and leaf-matter in one-month old compost.

It is important not to put too much soil when making compost: just sprinkle over green manure and form a thin layer of soil. Covering the compost with banana leaves is also very important to ensure the heap does not dry-out and allow to green manure and organic matter to de-compost. A wellconstructed compost heap will have its internal temperature rising above 40°C during de-composting. The height of compost should shrink down to one third of the original height after which the compost is ready to be used as an organic fertilizer. Application of compost is explained in the 'Further Explanation' chapter below

TRANSPLANTING

There are two method of planting a rice crop. One is by directly sowing the seeds into the soils; whilst the other is to raise the rice plants as seedlings and transplanting it onto the prepared land or garden site. When direct sowing, seeds are place into the soils at depth of 2 -3 cm; and when transplanting, rice seedlings are placed into the soil at depths ranging from 4-5 cm deep. In transplanted rice crop, there is uniformity in height, form, growth, and eventually tillering, flowing, and grain ripening. This will enable uniform harvesting of the rice crop. In directly seeded or sown crop, there may be high variability in crop establishment and the rice crop flowering, ripening and harvesting may not be uniform or within a narrow period to time. This handbook promotes transplanting of rice crop over direct-sowing. Direct sowing may be easier and a quicker way of planting one's rice garden but it does not guarantee good rice crop establishment. It produces an uneven crop germination due to different soil conditions, resulting in variation in the crop stand, and the flowering and grain ripening is spread over time., Transplanting is less easier when its technique is understood well and properly practiced, it can produce a more uniform crop establishment, even crop ripening and hence allowing only one-time harvesting. Transplanting of rice offers effective care and management on a small rice plant nursery bed then when directly sowing a large area of land where the crops' emergence may be problematic due to the sowing of seeds into the deep planting holes or some seeds sown are not viable or were bad seeds. It is better to obtain poor germination in the nursery bed rather poor emergence of rice crop in the field or garden because you can still make up for the short-fall in the nursery and in the field. A garden will fail plants growth or lines due to poor germination the crop is directly sown needs to be avoided at times. Sowing seeds in the nursery bed allows one to obtain the level of germination and number of seedlings required for planting the prepared field or garden. Whether one obtains 100% germination or less than 50% germination in the nursery, you can select healthy seedlings to cover the entire area of your garden land or field then to get disappointed when your directly sown field has many missing plants. It is preferable to over-sow a rice nursery seed-bed and selecting healthy and vigorous seedlings to transplant the entire garden or field, giving 100 per cent crop establishment. Transplanting of the garden or field will also allow quick and early establishment and a uniform crop growth and formation. This will then allow a more synchronized panicle initiation, flowering, and grain-filling. This will allow for a narrow period of crop performance, thus minimizing significant insect pests attack. Less or no pest damages means a good crop harvest is expected. In terms of timing, sowing and transplanting should take place when the wet season starts while harvesting should take place in a dry season. It takes about 130 and 136 days from the time of sowing to harvesting for late-maturing varieties, 113 and 125 days for medium-maturing varieties, and 110 days for early-maturing varieties. Based on the information of varieties, you should decide the appropriate timing for sowing.

Preparing Seeds

For Planting You need to have good rice seeds for successful rice cultivation. After the 1st season, you should keep and use the best seeds from your own garden to continue to cultivate rice. You should use the same variety of seeds for a garden to manage your farm well. If you don't have seeds, you can enquire about the available seeds which are resistant to pests and your environment at the local office of Division of Agriculture & Livestock (DAL) or National Agriculture Research Institute (NARI).

After obtaining seeds, you should check if your seeds are heavy, full and uniform in size. Make sure your seeds do not contain weeds or trash. If you find weeds or trash, you should winnow it (winnowing is explained in the Chapter 3.4 below). In order to ensure the quality of seeds, you should conduct germination test in the following manner. You can consider the seeds are in good quality if the germination rate is higher than 80%. (1) Select a number of small random samples from the seeds to be planted and select a subset (e.g., 100 seeds) of the combined sample. Soak the seeds in water for 24 hours; (2) Arrange 100 soaked seeds in a grid pattern on a wet paper towel (10 seeds by 10 seeds) and place a cover for protection; (3) Ensure paper remains moist; (4) Count the germinated seeds 3 to 5 days later and record the germination percentage; and (5) If more than 80 seeds have germinated, seeds have more than 80% of germination rate.

Making and Caring of Nursery Bed

You need to sow rice seeds in your nursery bed to grow seedlings. If you want to cultivate 0.5 kg of rice, you need the nursery bed of 2 m² size. Follow the steps below when making a nursery bed to raise rice seedling: (1) Soak rice seeds in a dish or container of water for three days in an open sun before sowing, change water every day; (2) Designate the area for the nursery bed and place logs to fence the nursery bed; (3) Put the banana leaves with some holes beneath the nursery bed to make it easier to transplant seedlings later without damaging the roots; (4) Put the soil over the banana leaves for less than 5 cm or a half finger height. It is good to use compost as the soil for nursery bed if it is ready; (5) Sow rice evenly on the nursery bed (give at least 1 cm space between the seeds); (6) Cover the seeds with light soil and sprinkle water over the nursery bed; (7) Cover the nursery bed with banana leaves to make uniform germination and for protection; (8) Sprinkle small amount of water every morning and afternoon; (9) Seeds should germinate in 3-5 days. Remove banana leaves to give sunlight after germination; and (10) Seedlings will be ready for transplant when they get three leaf-stage or two weeks after sowing;

Farmers sowing his rice seeds onto the specially prepared seed-bed using composted organic matter rich soils.

The resulting germinated rice seedlings.

Transplanting

Transplanting the seedlings from the nursery bed to the rice garden should take place when seedlings grow three leaf-stage. You have to be careful to transplant in the right time because if seedlings grow too big, the roots might be damaged when transplanting and will not grow further. It is better to transplant in the early morning or in the late afternoon when the sun light is not strong. Spacing between the seedlings depends on the variety (size) of rice, but should be around 20-30 cm x 20-30 cm (depending on the variety) to give enough space for people to pass through such as for weed and pest control.

The laying out of a garden field for transplanting the rice seedlings can be done using sisal/ agave strings.

The following are simple steps used in transplanting the rice seedlings:

(1) Monitor the growth of the seedlings and wait until it has reached the three leaf-stage or for two weeks after sowing; (2) Make holes by a stick along a rope as a guide to plant seedlings in rows. Give about 20-30 cm x 20-30 cm spacing; (3) Bring seedlings with full soil and place them in a reachable distance in the rice garden; (4) Select only good seedlings and transplant one by one (or two by two) immediately, covering the root with soil gently. Press soil around the seedling softly; (5) Go backward when planting not to damage the planted seedlings; and (6) Give small amount of water in the afternoon after transplanting. The roots are exposed outside the soil and the seedling may look weak. The roots need to take some water to recover.

Rice seedling transplanting hole are may be using stick picket, along the straightened strings.

Rice seedlings been picked out of the seedling bed.

Individual seedling is selected for transplanting.

Seedling being placed into the planting hole and soil pressed firmly around it.

Care and Management of Rice Garden

You have to regularly visit your garden and check the condition of the garden and take necessary and timely care. It is during these visits, that the farmer will need to do weeding, some soil tillage to improve the surface soils, and undertake some water control and pest control activities.

Weed Control, Soil Mounding and Soil Improvement

3 weeks to 1 month after transplanting, you need to control weeds at your garden. Weeds take nutrients from the soil and compete with rice growing. Weeds also attract pests and diseases as well. If you control weeds, you are also controlling pests and diseases as well. What you have to do is to remove weeds from roots using a hoe between rows and put them back to the soil as organic fertilizer as indicated in the picture below. Make sure you shave and stir more than 10 cm to remove roots completely and mix with soil well. At the same time, you should land up area where the soil flew out. You should also apply compost during the time of weed control. Just bring compost between the rows and stir together with existing weeds.

In terms of frequency, one time weed control 3 weeks to 1 month after transplanting is enough. You should do weed control, soil mounding and soil improvement all at once to save your time and energy. After 1 month, weeds rather work to prevent water evaporation of the soil.

Farmer weeding the rice garden using a weeding hoe.

Water Control

The ideal moisture content of the soil is 50-60%w.b. A good indicator of 50%w.b. moisture content in the soil is when you work the soil with your hand and are able to make a mass. A 60%w.b. moisture content or higher in the soil is indicated by you being about to able to squeeze small amount of water out of soil ball held in your hand.

Water Control

The ideal moisture content of the soil is 50-60%w.b. A good indicator of 50%w.b. moisture content in the soil is when you work the soil with your hand and are able to make a mass. A 60%w.b. moisture content or higher in the soil is indicated by you being about to able to squeeze small amount of water out of soil ball held in your hand.

Soil ball made between the palm-ofthe-hand with 50% moisture content.

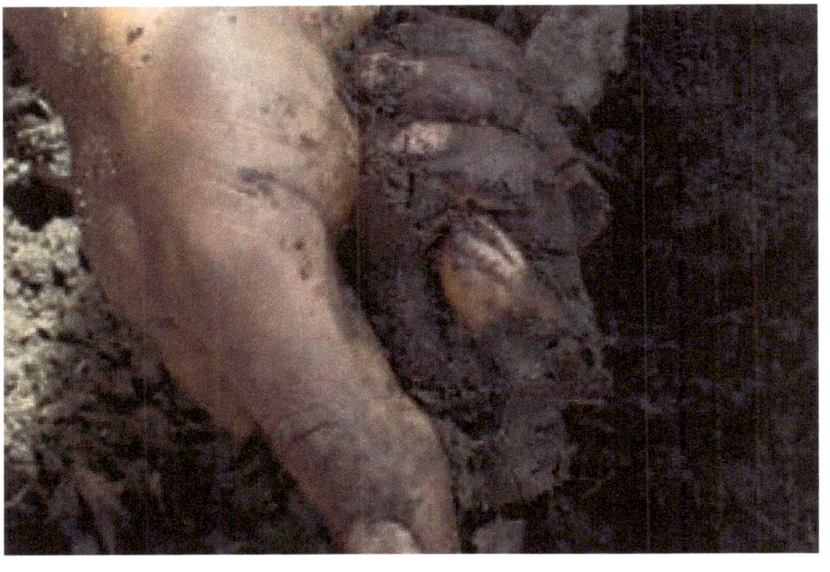

Soil with higher moisture content of above 60% may look this when worked between of palm-of-the-hand.

Check the moisture content especially when the weather changes and give water if the moisture content of soil is less than 50%w.b. This should be practiced in the case where irrigation is practiced. If the moisture content is more than 60%w.b., you have to open the gate of the trench and release excess water. When rice starts to bloom, it is better to create a dry environment soon after 4-5 days of flowering so then the rice crop will stop further tillering and panicle initiation and concentrates all its energy to filling in the grains. If you see some water in the trench or paddy field, improve the drainage to remove excess ground and surface water to influence the crop to start filling-in the grains rather than causing more tillering and more flowering, with extended period of flowering.

Pest Control

It is also important to protect your garden from pests. You have to go to the rice garden every day from the time of flowering stage till harvesting and control pests if you see any, before it gets too late to control. Some of the major pests and control measures are indicated below. (1) Stem Borer Stem bores are one of the major pests in Papua New Guinea. The life cycle of stem borers is 41-60 days. Eggs of stem borers are laid on leaves or leaf sheath. Larvae bore into stem, eat through the node, and bore down to the base of stem. Full-grown larvae pupate in rice straws or stubbles. Infested rice plants get white heads or white colored panicles. Boring causes panicles with half or empty grains, destroy the crop and reduce the yield. As a control measure, stems with white heads should be pulled off from stem based and opened up to reveal the larvae and pupae. Stem borers larvae and pupae must be pick-off by hand and destroyed immediately. Ants are one of the natural enemies of stem bores; once larvae are put down on the ground, ants will prey on them. Having a fallow or rice-free period for two months between cropping is the best preventative measure as this allows the removal of the habitats of stem borers. Harvesting all at once is important; if you leave some rice plants still standing after harvest, adult stem borers can continue to lay eggs on those plants, thus maintaining its life-cycle. Harvesting all at once needs to be applied to all the surrounding farms and destroying all standing crops after harvest. Just doing this to your farm is not enough as other farmers' field would provide the habitats to stem borers and other pests. Therefore, harvesting at the communities level at the same requires good cropping planning. This can be achieved by communities developing their own crop calendar to implement synchronous cropping, which is explained in chapter 4.1. It is also important to remove stubble and stems completely and immediately after harvesting not to give the habitats to stem borers. Stubbles and stems should be removed completely and ploughed into the soil.

White-head' is the sun-bleached rice plant's panicle that is an indication of rice stem borer attack.

Adult moth of the rice stem borer.

A stem boring larvae or worm extracted from a rice stem.

Rice-stem borers being devoured by ants.

(2) Rice Bugs Rice bug is one of the major pests in Papua New Guinea. The presence of the insects can be easily determined by an offensive smell it gives off when touched. Milky stage of the rice plant attracts rice bugs. They prefer to feed on the endosperm of the rice grain during the grains-filling (milky) stage and often cause empty or stained grains after the milky stage. The bugs feedings also causes the grains to loss form and becomes stained when milled. The bugs' feedings causes yield losses and lowers the quality of grains due to the presence of defective grains. The life cycle of rice bugs is 88-101 days. Rice bugs live off on wild hosts, mainly weeds of rice family (Poaceae) one or two generations before migrating into the rice field at the flowering stage. As a control measure, rice bugs should be removed by hand, using a net or plastic bag by sweeping the top of the panicles. Rice garden should be visited in the morning and afternoon during the milky stage to carry out pest scouting and administer control measures. As the preventive measures, weeds of grass and rice family should be weeded or slashed eliminate the habitats of rice bugs.

General appearance of the rice crop been attacked by rice bugs.

Adult rice bug, on the rice panicle with grains at the milky-stage.

A road side Grasses like this act as the alternative hosts for rice bugs.

Rice bugs being caught and collected in a sweeping-net.

(3) Brown Plant Hopper Rice is the only host plant for brown plant hoppers (BPH). The life cycle of brown plant hoppers is 19-24 days. Brown plant hoppers lay 200-500 eggs on the rice leaf-sheath and the nymphs and adults feed at the lower stems of rice plant. They migrate from other rice fields and can reproduce or generate 2-3 generations within on crop of rice. Hopper-burns observed in crops area main symptoms of serious BPH infestation, causing complete death of the plants, resulting in yield loss. They also transmit plant virus and plant diseases related to virus to healthy rice crop. As the control measure, BPH should be removed by net or plastic bug to reduce the number. However, once the brown plant hopper occurs in high density, no other option is available to control the BPH except for insecticide spray or early harvesting. As the preventive measures, you can use the rice variety which is resistant to the brown plant hopper such as NR 1. But any variety can become susceptible to BPH so it is advisable to check the latest information with DAL or NARI. Having a fallow or rice-free period for two months between cropping is the best control and preventative measure removing the habitats of BPH and stem borers. Harvesting the entire rice crop at once is important; if you leave some rice plants from the crop harvested still standing in the garden or field, all the BPH and stem-borers will move to those plants and can survive to cause further damage. Synchronous rice cropping is another control measure for avoiding the BPH population. It is thus important to remove stubble and stems completely and immediately after harvesting so as not to give BPH a habitat or host plant to thrive on. Stubbles and stems should be removed completely and ploughed into the soil.

Female BPH 5-6mm in length.

Male BPH 5-6mm in length.

Cluster of BPH infestation & causing hopper damage on rice.

(4) Rice Root Aphids Rice root aphids (Tetraneura) are pests found in Madang and East Sepik Provinces. They are also known as the grass root aphids because they infest the roots of other plants such as yam. They suck the plant sap from the roots and cause change of colour and distort the growth of rice grains. When these plants are pulled out, a large number of aphids can be seen on the roots. The control and preventive measures are now studied and yet to be made known.

Root aphids on rice plants roots.

HARVESTING AND SOIL TREATMENT AFTER HARVESTING

Harvesting

There are three basic method of determining the harvest time for your rice crop. One is by counting the number of days between planting and harvesting. Each rice variety has a crop-duration measured in the number days from planting to harvesting, when the optimum ripening age is reached and the crop can be harvested. This number varies from 110 to 135 days-after-planting. Another method for determining the harvest time is the percentage of grains that have turned yellow and are full at ripening. It is harvest time when about 80% of grains turned yellow regardless of the weather conditions. The ripening grains should never be allowed to fully ripen at 100% level because at this stage the rice grains and crop starts deteriorating and can be devour by birds and wild animals. The third method of determining the harvest time is by the grain's moisture content. The ideal grain moisture content for harvest is between 18 and 22%w.b. Grains should be firm but not crumbly when bite or press between the teeth. When harvesting, it is important to harvest all the grains together at once. If the crop is big and cannot be harvested in one day, you must organize a hired or pooled labour and harvest the crop in a day or two. Never leave your rice crop to another week or fortnight to harvest. Doing so will cause your crop to degenerate and allow insect pests, vermin and birds to thrive eating off your ripen grains. Leaving your crop to a delayed harvest may generate through crop lodging, grain shattering, and mouldy or pre-germination whilst still in the field. Therefore, the use of correct timing to harvest will help reduce crop losses as described above. Delayed harvesting will also cause grains to fluctuate in its moisture content over the period they are still in the field, casing grains to deteriorate in physical integrity. Harvesting too early will result in a larger percentage of unfilled or immature grains, which will lower the yield and cause higher grain breakage during milling. Generally the ideal harvest time lies between 130 and 136 days after sowing for late-maturing varieties, 113 and 125 days for medium-maturing varieties, and 110 days for early-maturing varieties. As tools for harvesting, you should have bush knife or sickle to cut the base of the stem. You should have hoes and shovels to treat soil immediately after harvesting. Before harvesting the entire crops, you have to harvest best rice seeds to use for next cropping. It should be harvested in the middle of garden because the rice at the edge of the garden could have been affected easily by surrounding environment and seeds might not be ideal. You have to select and harvest healthy and full mature rice panicle for seeds and separate from other grains. After harvesting best rice seed panicles, you should harvest all the rest of grains by cutting the base of the stems.

Training youth harvesting rice crop using the harvesting knives or sickles.

Soil Treatment Just After Harvesting

After completion of harvest, you have to remove all the weeds and stumps from the roots and plough them back into the soil by ploughing with hoes or any other gardening tools, to prepare for next cropping. Soil treatment should be done together with harvesting, to remove all the habitats of pests, put green manures back to the soil, and give enough time for green manure to de-compost for next cropping.

Soil tillage using hoe before planting and soon after harvesting.

3 Post-Harvesting Post harvesting is an important process to obtain and keep good quality of rice. The basic steps of post harvesting are indicated in the followings.

Selecting Best Seeds

You should select best seeds from the selected panicles as seeds for next cropping. Pick healthy, fully-matured, and uniform seeds in size especially from the center of the panicles. Try to select 40% of your selected panicles as seeds and the rest for usual consumption in order to obtain better germination ratio and uniform quality.

REFERENCES

1. Adachi K (1999) Isolation of hydrogenotrophic methanogenic archaea from a subtropical paddy field. FEMS Microbiology Ecology, 30: 77-85.
2. Adhya TK, Pattnaik P, Satpathy SN, Kumaraswamy S and Sethunathan, N (1997) Influence of phosphorus application on methane emission and production in flooded paddy soils. Soil Biology and Biochemistry, 30: 177-181.
3. Adhya TK, Rath AK, Gupta PK, Rao VR, Das SN, Parida KM, Parashar DC. And Sethunathan N (1994) Methane emission from flooded rice fields under irrigated conditions. Biology and Fertility of Soils, 18: 245-248.
4. Adhya TK, Bharati K, Mohanty SR, Ramakrishnan B, Rao VR, Sethunathan Nand Wassmann R (2000) Methane emission from rice fields at Cuttack, India. Nutrient Cycling in Agroecosystems, 58: 95-105.
5. Allen LH Jr, Albrecht SL, Colon-Guasp W, Covell SA, Baker JT, Pan D and Boote KJ (2003) Methane emissions of rice increased by elevated carbon dioxide and temperature. Journal of Environmental Quality, 32: 1978-1991.
6. Anastasi C, Dowding M and Simpson VJ (1992) Future ClLJ emissions from rice production. Journal of Geophysical Research, 97: 7521-7525.
7. Aulakh MS, Bodenbender J, Wassmann Rand Rennenberg H (2000a) Methane transport capacity of rice plants. I. Influence of methane concentration and growth stage analyzed with an automated measuring system. Nutrient Cycling in Agroecosystems, 58: 357-366.
8. Aulakh MS, Wassmann R, Rennenberg H and Fink S (2000b) Pattern and amount of aerenchyma relate to variable methane transport capacity of different rice cultivars. Plant Biology, 2: 182-194.

CHAPTER 8

Soil And Water Conservation Through Horticultural Intervension in Hilly Areas

Chapter Contents

- **Introduction**
- **Scenario of Northeastern Region**
- **Why Soil and Water Conservation**
- **Problems Related to Soil Erosion**
- **Development of Horticultural Land Uses**

INTRODUCTION

Land, which is the most precious heritage and the physical base of biomass production of life supporting systems, is finite. In this natural non-renewable endowment, the share of our country is fixed at about 329 million ha. It is not only inelastic but also heterogeneous in different parts and regions of the country with a definite set up, capabilities, suitability for different land resources. Conservation of land resources can promote sound land use to match with the land capabilities or suitability and to initiate correct land resources, development/suitability in the country. A close look at the present health of the soil and water resources reveals their wanton misuse and degraded environment. About 173 million ha covering slightly half of the country are threatened by various types of degradation like salinity, alkalinity, water logging, ravenous and gullied lands, areas under ravages of shifting cultivation, desertification, etc. About 800 ha of arable land are being lost annually due to ingress of ravines. There are specific problems of land degradation due to open-cast mining operations, using good productive land for brick kilns, coastal erosion and seawater ingress, excessive erosion and land slides in the crumbling hill areas. Our forests and grass lands have been over exploited. Frequent occurrences of floods and droughts in different parts of the country are evidence of improper land use in the catchments and inadequate conservation and use of rain water. The problem of land degradation has brought us face to face with the ever increasing depletion of the productivity and the basic land stock through nutrient deficiencies on the one hand and the ever growing demand for food, fodder, fibre, fuel, land based industrial raw materials and many non-farm land uses on the other hand.

SCENARIO OF NORTHEASTERN REGION

The north eastern region comprising of states of Arunachal Pradesh, Assam, Manipur, Meghalaya, Mizoram, Nagaland, Tripura and Sikkim is endowed with wide variety of flora and fauna. Degradation of land and water resources is a serious problem in this region. The adverse biotic impact has resulted in degradation of forests and this trend has been highlighted in many recent reports. Fast paced and multi-faceted development and ever increasing population have created tremendous pressure on land to provide basic requirement essential for survival. To meet these requirements, the limited natural resources are being over-exploited resulting in widespread eco-system degradation. The northeastern region is highly susceptible to acute soil erosion problems due to its undulating topography and high intensity rainfall. The primitive cultivation practices like jhum and bun further enhances these degenerative trends and rampant deforestation, wild fires, extensive grazing, unscientific mining and quarrying, etc., are adversely affecting the overall ecological condition of the region. Control efforts have not succeeded to desired scale. The authors strongly feel for the need of review and re-oriention of the planning process and programmes particularly for conservation and maintenance of natural resources such as soil, water, plants and animals. The climatic conditions and topography in different North East states varied from plain to high altitude thereby providing congenial altitudinal variations for the production of forest, cereal crops, vegetables, tropical and sub-tropical and temperate fruits. The economy of the states mainly depends on forest wealth, horticultural crops, agriculture and animal wealth. All these four components together contribute to the preservation of eco-system, soil conversation and healthy environment. The technical measures for conservation and development essentially are:

- Tree planting and reforestation
- Water harvesting
- Terrace and other engineering structures
- Agronomic, grassland and pastures development and other socio-economic measures for conservation and management of the soil.

It is evident that combating the degradation of our natural resources especially soil, water and vegetation and investing in their conservation for future generation will be a major practical task promoting sustainable development and nature protection. The application of modern techniques and development of new methodologies to cope with the widespread problem of soil degradation has become imperative to protect cultivable and uncultivable lands from the ravages of erosion, enhance and restore soil productivity, reverse degradative trends and restore degraded soil.

WHY SOIL AND WATER CONSERVATION

Land and water are natural resources that are essential for the existence of life and are the two variable factors for which management has become most essential. Land provides food, fuel, fodder and shelter besides supporting secondary and other economic life supporting system. However there has been a continuous depletion of land resources and the quality of land is deteriorating due to various factors like soil erosion caused mainly due to shifting cultivation, high rainfall, large scale deforestation, reckless mining activities, overgrazing, general mismanagement, etc. Such soil erosion leads to degradation of soils' physical property and loss of plant nutrients It takes nature 600-1000 years to build 2.5 cm of top soil but get displaced in a year only due to misuse. It has been reported that 6000 million tones of productive soil are lost every year from about 80 million hectare of cultivated land alone in India. It has also been proved that soil lost from unprotected land is about 120 tonnes /ha/yr and may go as high as 300 tonnes /ha/yr. Thus, a part from depletion of fertile soil erosion results in the loss of runoff water, plant nutrients and micro flora, siltation of reservoirs and riverbeds thereby adversely affecting irrigation and power potential, causing floods in plain and valley which damage crops, animals, habitation, communication, etc. But most of them adversely affect agricultural production, forest productivity and availability of water both for irrigation and drinking besides bringing about a disturbance in the soil and water balance

PROBLEMS RELATED TO SOIL EROSION

Normal erosion is a continuous process where soil is regenerated by natural means at the same rate as it is removed. But when protective vegetation is disturbed by cultivation, grazing or burning, the natural balance is upset and soil becomes exposed to most serious causes of erosion, i.e., water and wind. Under these conditions the soil can be washed away at a faster rate than it can regenerate resulting in a net loss of soil. It causes exposure of bed rocks and silting of rivers and dams. The main cause of soil degradation in the region is as follows.

- Shifting cultivation

It is known as jhuming and regarded as the step in transition from food gathering or hunting to food production. This traditional practice is still predominant in this region. In this system farmers burn the cleared vegetation and cultivate plots of land in the virgin forest until the yields of the crops fall below subsistence level. At this point the farmers abandon the land to natural fallow and move to a new site. When the original area is considered to have recovered fertility, the cultivators return to repeat the process, until once again the fertility of the land is apparently exhausted. As a result of population explosion, demand for food and fuel increased and land availability for agriculture has reduced. As a result, the jhum cycle of 10-15 years is reduced to 3-5 years. Earlier when jhum cycle was long the abandoned land got sufficient time for regeneration of forest but now due to reduced jhum cycle land does not get time for regeneration of vegetation. Indiscriminate felling of trees on the hill slopes brought an undesirable eco-imbalance. Further, the hill tops are the main source of water; deforestation of this hill top led to the elimination of water source. This, in fact, ended in the loss of top soil. Coupled with this, deforestation drastically reduced the retentive capacity of the soil. Erosion of soil in the catchment area resulted in silting of the reservoirs and streams leading to unprecedented floods. Hence, this situation needs to be tackled on top priority to keep the ecological imbalance in tact as well as to meet the fodder, food, fuel requirements, etc., in these states.

Soil Erosion Calendar of Shifting Cultivation System

Month	Agricultural operation	Erosion problem	Soil erosion (t/ha)	
			Min	Max
January to April	Selection of plot, forest cutting, burning and cleaning of hill slopes and sowing begins	Displacement of loose soil materials to down hills and rolling down of earthworm casting, soil erosion as above and wash due to rains.	0.0	22.4.
May	Sowing/weeding	Heavy soil wash, faint drilling at foot hills on silt deposits	0.2	61.9
June	Weeding	Heavy wash of soil aggregates	0.2	45.4
July	Weeding/harvesting begins	Heavy wash of soil aggregates, crop root exposed, farm soil visible	1.8	21.9
August	Harvesting and occasional weeding	Soil wash continues	1.0	29.6
September	Harvesting	Moss appears, soil erosion slows down	0.1	13.8
October	Harvesting	Soil erosion appreciably reduced	0.0	2.7
November	harvesting	No erosion, moss turns blackish	0.0	0.0
December	Harvesting/threshing/carry harvest back to home	No erosion	0.0	0.0
Year	Cropping with zero tillage on steep slope	Heavy soil wash	3.3	201.4

- Unscientific land use on hill slopes

The entire resource degradation process in the region is closely linked up with the land use system. Forestry is the most dominant land use system in the region followed by agriculture, horticulture, animal husbandry and non-agricultural uses (urbanization, commercial establishment, etc.). Evaluation of some of the land use systems practices in the region indicates that most of them are hazardous to resources and are not conducive to the aims of permanent agricultural systems with sustainable production. Horticultural crops grown on the hill slopes without proper soil and water conservation resulted in soil erosion. The soil erosion varied with the extent of disturbances caused to the soil surface. Colocasia, tapioca, sweet potato, turmeric and ginger are the crops, which resulted in movement of soil to the foot hills during the process of harvesting. Vegetable crops grown on the slope without proper soil and water conservation measures also resulted in this type of soil loss.

Soil erosion on hill slopes

Sl.No.	Land use/practice	Soil loss (t/ha/yr.)	Reference
1	"Bun" system for raising tuber crops	40-50	Singh(1970)
2	Pineapple along the slope	24-62.6	Ghosh (1976)

- Over exploitation of forest

Exploitation of forest indiscriminately increases the soil erosion on hills and flood in down stream areas. Fuel shortage becomes acute. This again means further encroachment of forest land resulting in more denudation, environmental degradation and loss of biological components of soil and vegetation. Thus a vicious circle may crop up. Even excessive grazing by cattle may also damage forest. The best example is Cherrapunji in Meghalaya, which has been famous till recently for recoding the highest rainfall in the world. The place is suffering from acute scarcity of drinking water. With an average rainfall of over 1150 cm or more one would expect Cherrapunji to be clothed in lush green forest. But what one sees now is a desertified barren bed along the slope. Heavy deforestation for augmenting fuel/fire need and harsh climate of 1150 cm of annual rainfall distributed to whole year has come down in 4-5 months during monsoon.

- Land degradation

The extent of land degradation that follows use of forest areas for agriculture is largely determined by the level of management. Apart from soil loss that accompanies land clearing and early stages of plantations, there is also severe nutrient loss. The practice of jhuming or shifting cultivation in northeast region has increased the problem of land degradation. The involvement of such a large area in shifting has caused large scale deforestation, soil erosion, loss of productivity, ecological imbalance and land degradation. The growing population pressure has resulted in misuse of land resource and national options for high value plantation crops have severely affected the tropical forests. Estimates have revealed that nearly 88.3 million tonnes of soil is lost annually as a result of shifting cultivation in NEH region. The degraded land as can not be put to any productive use.

Soil and Water Conservation Measures

The water conservation is inseparable from soil conservation. It is the application of all measures necessary to conserve the whole complex of land and water resources. The maximum benefit of a variety can be derived only after maintaining a high fertility level. Therefore, soil and water conservation measures should be primary objective of research programmes. There are two approaches that have been used to reclaim degraded soils and intensify agriculture production system, i.e., engineering approaches and ecological approaches. Mechanical soil and water conservation measures are required for controlling soil erosion, retaining maximum rainfall within the slope and safe disposal of excess run off from the top to the foot hills. These structures are used in case of extreme soil degradation, where other approaches are not possible or slow. The measures useful in controlling are discussed below. Contour bunds: Bunds are either mechanical or vegetative barrier created across the slope. The purpose is to divert the excess run-off during rain to the waterways and to retain eroded soil. These bunds on steep slopes are created by way of excavating parabolic channels (0.3 m top and 0.2 m deep) on contours and keeping the dug out soil in form of a bund at the lower edge of the channel. These bunds require care in maintenance during first 2 years. Our experience showed that vegetative barrier alone will not serve the purpose on steep slopes. The vertical interval of these bunds may vary from 0.5 to 5 m depending on the land use, soil depth and slope of the land. Bench terrace: Bench terraces are flat beds constructed across the hill slope; spaces between two contours are leveled by cut and fill method. In micro-watershed involving steep slopes, experiences showed that only at foot hills few benches may be constructed to produce food crops through intensive cropping. The vertical interval of such terraces should not increase more than one meter. Such measures can be adopted where the soil depth is more than 1.0 m. In case the entire hill slope is to be converted into benches, the construction should start from foot hill. Half moon terrace: These are level circular beds having 1 to 1.5 m diameter cut into half moon shape on the hill slopes. These beds are used for planting and maintaining sapling of fruits and fodder trees in horticulture/agro-forestry land uses. Grassed waterways: These are channels laid out preferably on natural drainage lines in the watershed. As far as possible, natural courses should be used without much disturbances for draining out the excess water. At appropriate locations, stilling basin (water pools) should be created with land use of earthen and boulder pitched bunds for temporary detention of run-off water. This structure will also serve the purpose of energy dissipation of flowing run-off water. Water harvesting ponds: Dug out cum embankment type of water harvesting structure can be used for creating seasonal and perennial ponds at the foot of the micro-watershed for irrigation and fish farming purposes. The above soil and waster conservation measures can be created with the use of local resources. On hill slopes, soil and water conservation measures may be followed with horticulture land use system. Other steps required are: Contour bund at 1.0 m to 5 m. vertical interval in all the land uses with common grassed waterways. Bench terraces towards the foot hills for growing vegetable crops (Vertical interval should not increase over 1.0 m.) Half-moons terrace (1 to 1.5 m dia.) for fruit crops. Water storage at appropriate location for irrigation/fish purposes.

Watershed Based Land Use System

Watershed management is the development and management of the watershed resources in such a manner so as to achieve optimum production, which can be sustained without causing deterioration in the resources base or disturbing the ecological balance.

Principles of Watershed Management

Our research experiences proved that with the watershed based concept, it is possible to manage the hill slopes effectively while utilizing local natural resources. A high priority within watershed management practices has to be given to water conservation and water harvesting so that the farmer is attracted to the high yield enabled by increased water availability. Land has to be developed to receive the rainwater in such a way so as to utilize the maximum for plant growth and lead the rest into storage either surface water storage or ground water storage for later use by man and animals. The daily needs of the people for food, fuel, fibre, fodder and employment have to be adequately met before any other goals of watershed management.

DEVELOPMENT OF HORTICULTURAL LAND USES

In this system mixed horticultural crops including fruit, vegetable, root crops, etc., may be grown under the best management practices to keep the soil covered. The horticultural crop residue, i.e., stubble, leaves, etc., not only reduce the soil erosion but also serve as a mulching material and add organic matter to the soil. The rotation of annual horticultural crops in between the perennial trees is done to increase cropping intensity as well as maintain soil fertility, etc. Bunds, half moon terraces and grassed waterways are the major soil and water conservation measures required for land development. However, if vegetable crops are to be grown, then only bench terraces may be required on foot hills at a vertical interval of 0.5 to 0.75 m. The different horticultural land uses are to be formulated in the following pattern:

1. Agri-horticulture

In this system the 2/3 area (upper side) is covered under horticultural crops for which half moon terraces and contour bunds are prepared on the hill slope and 1/3 area towards down side is used for the cultivation of cereals, oil crops, etc., on the bench terraces. In this land use pattern, the following crops may be grown after the land preparation.

1) Fruit trees in half-moon terraces (triangular system of planting) on contour 2) On the contour bunds the pineapple in two rows should be planted at closer distance, which helps in soil erosion from contour area. 3) The interspaces in the contour are utilized for the cultivation of the vegetables. The legume vegetables like bean, cowpea, guar, pea and good over crop like sweet potato should be cultivated. 4) Ginger and turmeric can be grown in the interspace area in the contours.

2. Agri-horti-silvi -pastoral (model land use)

In this system the middle 1/3 area of the hills is taken for the cultivation of horticultural crops and upper 1/3 area and lower 1/3 area are being cultivated for establishment of economic forest plant's plantation with fodder and cereals, millets, etc., respectively. The middle portion is converted under contours and the fruit plants are planted in half moon terraces on the contours. The contour bund is utilized for pineapple planting. The two or three separate blocks of each fruit crop may be made so that cultural operations may become easier. The vegetables, root crops, rhizomatous crops, etc., are cultivated in the interspaces of the contour. The lower one or two contours may be used for pure vegetable cultivation.

3. Mixed horticultural land use

If the farmer is interested to grow only different horticultural crops in the land available with him the whole jhum land may be divided in the following pattern for developing mixed horticultural land use. a) 2/3 area from top towards lower hillside is converted into contour and 3-4 fruit blocks can be developed, i.e., banana block, orange block, lemon block, papaya block. The lower 3/4 contour is utilized for the cultivation of rhizomatous crops. After that 3-4 terraces may be completely utilized for the pure vegetable cultivation. The contour bunds are utilized for the planting of pineapple.

4. Horticultural land use (with fruit crops only)

The land use under pure horticultural orcharding system has high potentiality in the region. In case more area (jhum land) is available at a place in the selected site and the owners are too interested for growing of fruit trees collectively as co-operative farming type in order to bring their large area under horticulture then pure orcharding can be done and land use pattern may be developed accordingly, keeping in view about the soil and water conservation aspects. This system of cultivating the land will be highly profitable in the long run and area will be developed as fruit growing belt. The water and soil management practices are to be followed in a systematic manner so that the soil loss can be checked to a considerable stage. The following practices (management) are to be adopted while considering this land use. 1) The fruit plants like orange, banana, etc., are to be planted either in half moon terraces or in contour bunds. 2) If the slope is below than 25.30o the intercropping is to be practiced for getting the subsidiary income to the farmers and the four rows planting of pineapple after 10 rows of fruit trees across the slope will be advisable in order to check the soil erosion. 3) The legume vegetable should be considered for the cultivation as intercrop so that soil fertility may be enhanced. 4) If planting is done only in half moon terraces the chopping of weeds in interspace areas is advisable and the filler crops should be taken.

5. Horti-silvi-pastoral system

The horti-silvi-pastoral system has great potential to provide a sustainable land use system, which would maintain an acceptable level of production of fruits, vegetables, fuel wood, timber, fodder, etc., and at the same time, conserve the basic resources (mainly soil) on which production depends. This system was found economically viable and socially acceptable alternative to jhuming in this region.

6. Multi-tier horticultural system

1. Horti-horti three-tier system: areca nut + black pepper +ginger/ turmeric/pineapple/ Assam lemon 2. Silvi-horti-three tier system: MPT + black pepper + ginger/turmeric/ pineapple 3. Silvi-horti-two tier system (Parkia and pineapple or subabool and pineapple) a) Alder based farming system of Nagaland (alder and vegetables like potato, cole crops or alder and cereals like maize, rice, etc.) b) Alder based large cardamom system of Sikkim c) MPT + Assam lemon

7. Multi-tier system for plantation crop

Tea plantations in the region including Sikkim and Darjeeling play major role in the economy and employment generation. It has been established that tea and coffee plantations require sparse shade and Albegia, Dalbergia, Accasia have been used as the major tree species for the purpose, which in general formed a two-tier system of silvi-horticulture. Recently, black pepper has been introduced in some of the plantation for making the system more profitable.

For two-tier system

- Tree spp.(Albegia) + tea or coffee plantation
- Tree spp. (Dalbergia) + tea or coffee plantation
- Tree spp. (Acasia) + tea or coffee plantation

For three-tier system:

- Tree spp. (Albegia) + black pepper + tea or coffee plantation
- Tree spp. (Dalbergia) + black pepper + tea or coffee plantation
- Tree spp. (Acasia) + black pepper + tea or coffee plantation

8. Homestead gardening

A number of horticultural crops like guava, citrus, orange, banana, peach, pear, drum stick, etc., are grown with under story crops such as tapioca, colocasia, sweet potato, cucurbits, ginger, turmeric, beans, root crops, leafy vegetables, etc., having variations in their combinations from house to house depending upon area available, site condition, climate and choice of individual family. They are mainly grown for home consumption and surplus if any sold in the local market. Home gardening of temperate/subtropical zone Peach, pear, plum, apple, other indigenous fruits + cole crops, radish, carrot, turnip, pea and other indigenous vegetables. Home gardening of tropical/subtropical zone Citrus, guava, jack fruit, mango, coconut, areca nut papaya, pine apple, low chilling peach, pear and other indigenous minor fruits + okra, solanaceous vegetables, cucurbits, tapioca, ginger, turmeric, colocasia, sweet potato, black pepper and other indigenous vegetables along with water bodies (pokhari).

Selection of Crops and Varieties

Suitability of crops depends upon the altitude, soil and climatic conditions. Say, as an example, Manipur state can roughly be divided into 4 land zones viz. i) High hills: 900-2000m above MSL (apple, peach, pear, plum, apricot, potato, cabbage, cauliflower, radish, beans, etc.) ii) Mid hills: Below 500 m (citrus, banana, pineapple, papaya, guava, ginger, turmeric, chilli, brinjal, tomato, bean, sweet potato, tapioca, colocasia, etc.) iii) Foot hills: Bordering areas of hills (jackfruit, areca nut, black pepper etc.) Vast areas of the hills are suitable for cultivation of tropical, sub-tropical and temperate fruits viz.

Tropical	Cashewnut, banana, papaya
Sub-tropical fruits	Pineapple, citrus, guava, banana, gooseberry, etc.
Temperate	Apple, Peach, pear and plum
Spices	Chilli, turmeric, ginger, garlic
Vegetables	Tomato, chillies, brinjal, potato, radish, pea, colocasia, okra, pumpkin, bottle gourd, cucumber, carrot, cabbage, cauliflower, knol khol, french bean, winged bean, dolichos bean.
Tuber crops	Tapioca, sweet potato
Tree vegetables	Tree bean, tree tomato and drum stick
Suitable grasses for risers	*Stylosanthes guyanansis*, *Stylosanthes hamata* and Thinnapier, NB-21.

Crop Planning and Production Technology

The lower one-third area should be terraced and utilised for growing vegetable/spices or tuber crops. While planning crops, it should be necessary to keep in mind the requirement of farmers. Two crops round the year can be grown even under rainfed conditions. High yielding varieties of crops should be introduced. Usually the terrace risers have to be protected by growing some fodder grasses, so that apart from protecting the risers from erosion, it provides a subsidiary source of income to the farmers. Upper, the mid-one third, area should be utilized for fruit crops and the remaining upper one-third under the reserve forest (fodder or fuel trees). The plants should be planted in half moon terraces.

Other Methods of Efforts to Check Degradation

In the second half of 20th century protection efforts for forest resources and the wild life have been brought to the notice through several activities in whole country like 'Chipako Movement in UP/UA', 'Silent Valley Movement' in Kerala, 'Thingtam' in Mizoram, etc. The rich variety of vegetation in the Mawsymai sacred grove, near Cherrapunjee (or Sohra, as it is locally known), stands out as a vivid reminder of what the area was once like.

Effective Traditional Methods

Most of the agricultural activities in the region contribute considerably towards land and water degradation. Bench terrace cultivation in certain areas of Nagaland, Sikkim and Manipur, bamboo drip irrigation in Jowai district of Meghalaya, efficient water management system of Apatani plateau of Arunachal Pradesh, 'Zabo farming system' of Nagaland, high altitude farming of Buddhist Monpas of Kameng district of Arunachal Pradesh are examples of excellent management of biophysical resources based on local skill.

The deep-rooted traditional agricultural techniques cannot be washed away overnight. They can, however, be spontaneously modified to minimize the ecological ill effects. Jhum is an ancient method of cultivation wherein a patch of forestland is cut and burnt down to be cultivated. Bun is different in the sense that the slashed biomass is not burnt but ploughed in for organic content. These methods were relevant when the population pressure was less and the land was re-harnessed after 30 odd years. Today, the population pressure is such that cultivators come back to used land after three or five years. This is not sustainable, and that is why they do not get proper agricultural yield. For sustainable farming, agri-horti-silvi-pastoral system is one of the best land-use systems with adequate advantage of soil and water conservation. It has been proved that such a system can reduce soil erosion from 42 tonne/ha to 1.5 tonne/ha, besides providing for rainwater harvesting. This water can be harnessed both for pisciculture as well as irrigation. The commercial agriculture and the Green Revolution could not touch the NE. This was mainly due to the region's inaccessibility. Keeping this in mind, we should refocus our strategy and make it two pronged. One, augment food production through intensive highinput agriculture in the valley land and, two, convert rest of the areas into organic farming zones. Within the next decade and a half, the NE will not only be self-sufficient but will also be food surplus. For this, rainwater needs to be harvested. That may bring in more area under settled cultivation. To usher in a revolutionary change, all NE agriculturists should protect present resources and judiciously utilise the natural ones. Studies on alternative farming system to shifting cultivation indicate that agriculture with bench terrace and contour bunds as conservation base can provide stable alternate to switch over from shifting to permanent agriculture system provided maintenance of conservation measure is properly done. Agro-horti system of land use with subsidiary source of income through live stock rearing provides most favourable indication in favour of adopting mixed land use system as an alternative to shifting cultivation on steep hill sides. Such a system will certainly be technologically feasible, sociologically acceptable, ecologically sound and economically viable.

REFERENCES

1. Borthakur, D.N., Prasad, R.N., Ghosh, S.P., Singh, A., Awasthi, R.P., Rai, R.N., Varma, A., Datta, H.H., Sachan, J.N. and Singh, M.D. 1979. Agro forestry based farming system as an alternative to Jhuming. Reprinted from Seminar Proceedings on Agro-forestry organised by ICAR at Imphal, Manipur, May 16-17, 1979. 32p.
2. Ghosh, S.P. 1976. Development of suitable planting technique of pineapple in hill slope. Annual Report of ICAR Research Complex for NEH Region, Shillong.
3. Ghosh, S.P. and Ghosh, S. 1982. Water harvesting for fruit orchards in Dehradun valley. Proceddings: International Symposium on Hydrological Aspect of Mountainous Watersheds, November 4-6 Manglic Prakashan Saharanpur, U.P India viii-36 to viii- 41.
4. Singh, A. and Singh, M.D. 1981. Soil erosion hazards in North Eastern Hill region. Research bulletin No.10.ICAR Complex, Shillong.
5. Verma, A.K. 1998. Watershed management - A strategic approach for development of agriculture in NEH Region.Indian Journal of Hill Farming.11(1&2):1-6.

Index